THE HAPPINESS EFFECT

THE HAPPINESS EFFECT

HOW SOCIAL MEDIA IS DRIVING A GENERATION TO APPEAR PERFECT AT ANY COST

DONNA FREITAS

FOREWORD BY CHRISTIAN SMITH

OXFORD
UNIVERSITY PRESS

OXFORD
UNIVERSITY PRESS

Oxford University Press is a department of the University of Oxford. It furthers the University's objective of excellence in research, scholarship, and education by publishing worldwide. Oxford is a registered trade mark of Oxford University Press in the UK and certain other countries.

Published in the United States of America by Oxford University Press
198 Madison Avenue, New York, NY 10016, United States of America.

Library of Congress Cataloging-in-Publication Data
Names: Freitas, Donna, author.
Title: The happiness effect : how social media is driving a generation to appear perfect at any cost / Donna Freitas ; foreword by Christian Smith.
Description: Oxford ; New York : Oxford University Press, [2016] | Includes bibliographical references and index.
Identifiers: LCCN 2016009997 (print) | LCCN 2016016495 (ebook) |
ISBN 9780190239855 (cloth : alk. paper) | ISBN 9780190239862 (updf) |
ISBN 9780190239879 (epub)
Subjects: LCSH: Internet and youth. | Social media—Psychological aspects. |
Youth—Social life and customs.
Classification: LCC HQ799.2.I5 F745 2016 (print) | LCC HQ799.2.I5 (ebook) |
DDC 004.67/80835—dc23
LC record available at https://lccn.loc.gov/2016009997

9 8 7 6 5 4 3 2
Printed by Sheridan Books, Inc., United States of America

This book is dedicated to Katie, Andleeb, Ozzie, Cristianna, and Dion, the extraordinary students in my Hofstra memoir seminar. You struggled together to think about the biggest and most important of life's questions with openness and honesty, and allowed me to be present for this. You made me laugh and reflect and wonder at my luck in having you all in class. I am so proud of everything that you've become.

CONTENTS

FOREWORD

Margaret has to avoid Facebook because seeing how happy everyone else appears online makes her unhappy by comparison. Rob gets a call from a friend asking him to "Like" his new Facebook photo to save him from the possibility of it not being liked enough. Michael felt lonely because he spent most of high school trying to impress people on social media rather than spending time with his friends. These are just a few of the people you will meet in this book. And it is tempting to ask: What's the matter with kids today?

Everybody knows that the digital communications revolution—the Internet, social media, smartphones, online dating, and more—is transforming our society. But nobody really knows yet how these technological innovations are changing us and our ways of life—possibly including our very sense of self—and just how far it will go. We have lived long enough with this revolution by now to know that it *is* truly revolutionary, not a superficial phase. But we have not lived with it long enough to know what it really means for us in the long run.

Everywhere I travel to speak about the lives of youth, I am asked how the Internet, and social media, and smartphones are changing young people's lives—usually, it is suspected, for the worse. Is it making them less interested in real, face-to-face relationships? Is it turning them into self-centered egomaniacs? Is it causing them to disconnect from social institutions, like sports teams and churches? Is it distorting their sense of morality? As a sociologist, ever attentive to data, I have always had to answer that I don't really know, because there wasn't enough good research available. I think the changes *are* real and big, I would say, but beyond that I could only speculate. My answers were always incomplete and disappointing.

You now have in your hands a landmark book that answers these questions. Donna Freitas's *The Happiness Effect* provides the first really serious and reliable answers to these kinds of questions that parents ask every day. As a researcher, I am very excited about it, even as I find it troubling as a parent. Unlike a lot of writing in this area, this book is neither speculation nor sensationalism. It is serious, focused on a hugely important issue, and based on rock-solid empirical evidence. Freitas elegantly interprets the data—mostly by allowing young people to speak for themselves—in clear and accessible prose that is rare among academic writers. It deserves and needs to be widely read.

One of the most important findings in this book, to my mind, is the schizophrenic effect social media has on people's sense of self. Social media produces a world in which the problems and blemishes of real life are hidden behind virtual presentations of self that struggle, often obsessively, to be "Liked." One must always appear attractive, happy, and clever. And, as Freitas deftly shows, even while many users grasp the dehumanizing forces at work here, they find it difficult to keep themselves from playing into this virtual world's insidious grasp on human insecurities and fears. The damage is perpetrated mostly by the same people who suffer them. It is troubling to anyone who wishes to see young people growing up to be authentically secure, happy, realistic, and genuinely caring about the real needs of other people.

The pages that follow skillfully reveal the sometimes subtle, sometimes blatant ways that social media twist and distort young people's senses of self. I have seen hints of this in my own research, but this book nails it with force and insight. When I interviewed young people, smartphones and Facebook continually interrupted, metaphorically and sometimes literally. I knew social media was an essential topic. This book, which brings us inside the intimate thoughts and feelings of youth struggling to develop authentic senses of themselves, yet also wrestling to negotiate the immense pressures that social media places on them, provides answers to longstanding questions.

As to the author, Donna Freitas is one of the most qualified—perhaps *the* most qualified—scholar in the nation to write this book. She has spent years traveling all over the United States talking directly with many hundreds of students at every kind of college and university, about their experiences online, their personal and social identities, relationships, intimate emotions, sexual histories, views of their own generation, and much more. Freitas has profoundly and personally immersed herself in the worlds of which she writes and speaks—an up-close and labor-intensive research method for which there is no substitute. That gives her an unrivaled authority to write about these matters, one that deserves our hearing. Freitas is also an immensely talented interviewer, speaker, and writer. And, while she is adept at writing for and speaking to popular audiences, Freitas is a serious scholar who is careful about research methodology, data collection, and analysis. All of this makes her a rare combination of talents.

The book itself is also lucid and engaging—a real page-turner. Readers never have to trudge through difficult passages, numbers, or theories. The fascinating voices of distinct college students tell the stories for themselves. Yet those voices, in Freitas' skilled hands, add up to an exposition of important themes, insights, and conclusions that beg for our attention and response. The subject matter here is massively important, the engagement of it clear and captivating.

All of the above means that any number of people really *must* read this book, think hard about it, talk widely about its findings, and work on constructive responses: every parent and grandparent of children, teenagers, and twenty-somethings; every college and university student and administrator, student affairs worker, and faculty member; every middle and high school teacher and principal; every coach, youth pastor, and other youth worker who deals with teenagers and emerging adults; every young American who is not in college but in the work force or unemployed; every aunt, uncle, and mentor of youth; and every other person who cares about human beings benefitting from technological developments rather than being damaged by them. If you have a Facebook page, this book has much to teach you.

With this book now before us, we must all ask: are we capable of a response equal in seriousness to the massive consequences of the digital communications revolution that confront us?

Christian Smith
University of Notre Dame

PREFACE

TINY MEMOIRS

> *I think social media is kind of like a book cover: it*
> *can show what you are, but then also people*
> *can cover up themselves of what they are. Make*
> *themselves look better.*
>
> *Jason, junior, public university*

TRYING SO HARD TO FORGET OURSELVES

In a seminar on memoir that I taught at Hofstra University's Honors College, my five intellectually gifted and academically driven students contemplated falling madly in love while reading Patti Smith's *Just Kids*, what it's like to be young and Muslim via Eboo Patel's *Acts of Faith*, and the trials of grief in Joan Didion's *Year of Magical Thinking*. But there was something about memoir itself, about sitting down to contemplate life's meaning and purpose, that caused my students to question absolutely everything: their majors, their career paths, their backgrounds, the pursuit of true love. It pulled them up and out of their comfort zones and had them pouring out their deepest feelings. They talked endlessly of the overwhelming number of time-consuming commitments that ruled their days and nights and about how, when they stopped and took time to ask themselves why they were doing what they were doing, they weren't sure how to answer that question. It disturbed them.

There are so many things my students discussed that have stayed with me since that spring semester full of snowstorms and too-late spring blooms, but one in particular has played out in my mind over and over as I've done the research for this book.

One afternoon, after a particularly heavy snow, the students began to talk about their inability to sit still and their fear of doing so. *Fear.* This was the word they used. Being bored, having nothing to do, simply stopping and not using their phones to fill the silence. It scared them because their thoughts scared them. The way their thoughts about anything—life, relationships, love, work, school, family, friends, choices, their futures—would just *show up* in a way they couldn't control or block was very upsetting.

"I was just sitting there in my dorm room yesterday," one of the students was saying, "watching as the snowflakes were hitting the window and sliding down the glass." This student noticed how beautiful the snow was, they way it piled up and crystallized. He also noticed how peaceful he felt while watching it. But then, suddenly, in the silence and the stillness, other thoughts began to intrude, bigger thoughts about his life and what he was doing with it or, even more worrisome, what he was *not* doing with it. He began to feel a conflict as the uncomfortable collided with the beautiful and the serene. He was upset about his impulse to suppress these thoughts by grabbing his phone, because if he'd been on his phone he might never have had the discomfiting thoughts. But then again, if he'd picked up his phone, he would have missed this moment of beauty.

This was what upset him most of all: the possibility of missing the moment of beauty.

The young man went on to tell the class how he couldn't remember the last time he'd just enjoyed a small bit of beauty, let himself be taken by it. His *go, go, go, do, do, do* schedule and nearly constant résumé building and social media updating had robbed him of those moments. But his fear of the thoughts that followed, his inability to handle them, was behind his motivation to never let himself pause and open himself to the possibility of catching more of them.

How do we deal with this conflict? he wondered.

Should thinking be so scary?, the class asked collectively. Shouldn't we be able to pause and live in the moment? Wasn't this an important thing to be able to do? If we aren't able to stop and think, will we ever be able to pursue the things that mean the most in life, or even know what might make our lives more meaningful? What if life becomes an endless cycle of escaping difficult thoughts?

Thoughts and feelings, the kinds of things that come up when we just sit and contemplate the world and our lives, can make us feel vulnerable. And while vulnerability can be uncomfortable, it is often the very thing that leads us to ask the important questions, the big ones that going to college and growing up are supposed to be about. On social media and via our devices, we are learning to shut out these things, these moments, these feelings.

We are putting up a shield around our own vulnerability.

I wasn't too far into this research before I knew that this study I'd begun on social media was really about happiness, about how young adults are learning they must appear happy at all times, presenting to the world what looks like the perfect life. Yet in always trying to appear happy, perfect, even inspiring and certainly enviable, we often neglect the very parts of ourselves that bring us true happiness, joy, connection, love, and pleasure. We become afraid of our true selves, of expressing who we really are, with all our flaws and imperfections. We begin to cover ourselves up, to clothe ourselves in words and images that mask the emotions and even the joys that define our hearts and minds and souls because they seem too intimate, and this intimacy seems inappropriate. We become good at hiding, we learn to excel at it, and society rewards us for the walls we've constructed with "likes" and "shares" and retweets and, ultimately, as young adults are learning so well, college acceptance letters and job interviews. By putting up these facades, by convincing our "audiences" not only that all is well but that all is *always* well, we sacrifice ourselves. By doing such a good job of "appearing happy," we risk losing the very things that make us happy.

When my student spoke of watching the snow fall against his window that day, of reveling in its beauty and then recoiling from what this moment of stillness evoked in his mind, he was really talking about recoiling from himself. Our devices and our compulsive posting and checking are helping us to flee ourselves. We are become masters of filtering away the bad and the sad and the negative. But in our attempts to polish away those imperfections and "put on a happy face," as one student told me, as we try to forget the darker and more tender sides of our humanity, we also risk losing the best parts of who we are.

WE ARE WORTHY

When I started mentioning the idea of a "happiness effect" to my friends and colleagues, about how I worry that it is costing us our humanity, our authenticity, and the things that make our lives meaningful, everyone told me that I *must* watch Brené Brown's TED Talks. I hemmed and hawed for a while, then finally sat down and watched the first one, on vulnerability, moving quickly on to the second, about shame. I found myself crying as I listened to Brown speak so eloquently about how in our imperfections we find our own worthiness and are able to encounter love and belonging, and how, in order to live wholeheartedly, vulnerability is essential. To experience true belonging and connection, we must be able to own our imperfections and our messiness.

All the people who told me to watch Brown's talks were right—I needed to see them. In the one on vulnerability, Brown talks about how in our attempts to polish our images, to appear as if we have it all together, as if everything is in order and our lives are always great, we end up turning away from the very things that can make us whole. She speaks of how much we lose when we attempt to make ourselves invulnerable and how what we lose, really, is *everything*—love, belonging, the things that make us feel the most worthy.

Brown asks us to "let ourselves be seen," "to love with our whole hearts, even though there is no guarantee," "to practice gratitude and joy in those moments of terror," and "to believe that we are enough." To achieve connection, the thing that makes life worth living, we must strive to do all of these.

Yet the public nature of social media teaches us to strive for the opposite. Social media and the ways we are learning to navigate it, as well as the ways we teach our children and students to navigate it, go against everything Brown talks about that makes life meaningful. In our attempts to appear happy, to distract ourselves from our deeper, sometimes darker thoughts, we experience the opposite effect. In trying to always appear happy, we rob ourselves of joy. And after talking to nearly two hundred college students and surveying more than eight hundred, I worry that social media is teaching us that we are not worthy. That it has us living in a perpetual and compulsive loop of such feedback. That in our constant attempts to edit out our imperfections for massive public viewing, we are losing sight of the things that ground our life in connection and love, in meaning and relationships.

Our brave faces are draining us. We're losing sight of our authentic selves.

WHAT TO EXPECT FROM THIS BOOK

When I set out to talk to college students about social media, I had no idea I'd end up writing so extensively about happiness. There was only one direct question in the interviews about happiness, yet happiness and everything related to it—being positive, hopeful, and even inspiring—came up all over the place in students' answers, no matter what the question. Happiness, at least the appearance of it, was a huge concern to them, as were the ills that stem from everyone's attempts to display happiness online, especially when those attempts ring hollow. The appearance of happiness has become so prized in our culture that it takes precedence over a person's actual happiness. By the time our children reach college, they know that a large part of their job is to present a happy face to the world, as they once might have presented a book report in class or performed a role in a school play. And we (the parents, teachers, coaches, and mentors in their lives) have helped push them to this place.

I don't believe these concerns are restricted to college students.

When I talk about the findings of this research with friends and colleagues, the conversation often sparks people to share their own struggles with online appearances versus the realities they live and experience. It is my hope that the themes addressed in this book will resonate with people far beyond college campuses. Everyone seems concerned with happiness, and never more so than when we attempt to paint our own portraits of it on our social media profiles.

There are many benefits to social media, primary among them (according to student participants in this research) the ability to connect with friends and loved ones who are far away. It is a basic tool for making real-life plans. Social media can allow us to be playful, expressive, poetic, flirty, and even silly. But it is clear that image-consciousness and professional concerns among young adults are eclipsing these benefits.

As a professor and teacher of many years, and as a former professional in Student Affairs for many years, I want the young adults I work with to feel empowered with respect to the things that influence their lives, choices, relationships, and behavior. I want them to become good critical thinkers about these things because critical analysis helps us to have power over these forces rather than being swept up in the tide. There is no doubt that social media is a major influence in young people's lives. Social media is also one of the newest, fastest-changing influences for all of us. It's still so nascent that most of us are reeling in the face of it, and only recently have we, as a society, begun to unpack how it is changing and affecting us for good and for ill.

Many people and organizations seek to offer up-to-the-minute accounts of the latest social media trends. Their efforts are laudable, and I draw on some of their research, but my intent here is different.

My priority is to showcase the voices and stories so generously offered by the students I interviewed and surveyed so that their college peers, their younger siblings, and the adult mentors and parents in their lives will have the opportunity to continue these conversations at home, in residence halls, in the classroom, and in the workplace. My hope is that the young adults who read these pages find themselves, their friends, and their peers represented in a way that empowers discussions of the issues they find most relevant. I hope that faculty and university administrators as well as high school teachers and administrators find these voices and stories useful in their attempts to talk about these important themes both inside the classroom and out. I hope that adults will find themselves, and their own struggles and thrills around social media, represented in these pages. I also hope that parents will come away from this book with a greater understanding of some of the challenges their children might face as they shift from high school to college while engaging in their near-constant public lives on social media. The students represented here demand our attention and call us to the conversation—and it will be a complicated one.

Listening to other people's stories can make us feel vulnerable. Stories open us up and empower us to talk about things we normally might never tell another soul. I think that's why, whenever I teach my memoir class, it always ends up hitting students at a deep level. When we are in the presence of others who have shown us their best and also their worst, their successes and failures, their joys and their deepest moments of pain, their stories call forth the same in us, they stir up the good and the bad, they ask us to look at love and loss, its presence and absence, our greatest triumphs and our darkest moments. They humanize us and call forth empathy.

The stories that follow are funny, sad, shocking, beautiful, frightening, insightful, and very, very real. I loved hearing what the students had to say, and I hope you love reading about them. They are the heart and soul of my research and the heart and soul of this book. I hope they open you up in thought-provoking ways that spark lots of conversations and big questions, like the memoirs that work their way into the hearts and minds of my students.

The stories presented here, I think, are like tiny memoirs of their own.

THE HAPPINESS EFFECT

INTRODUCTION

MASTERS OF HAPPINESS

I think that people want to show other people that they're happy. It's like a happiness competition sometimes, which is funny because I think if you're really happy, you don't feel that need to show other people that you're having a good time.

Blake, senior, private-secular university

It's kind of like how everybody says with their high school reunion, they want to go back and show off how great their life is. It's like that now, but you don't have to wait for your ten-year reunion. It's like that every day.

Brandy, junior, private-secular university

EMMA: THE WORST VERSION OF ME VERSUS THE BEST VERSION OF EVERYONE ELSE

Early on a sunny Saturday morning, Emma shuffles into the interview room. She has just rolled out of bed after a night of partying. Her hair is pulled back into a high ponytail, and she is adorned, head to toe, in sorority attire, T-shirt and sweats bearing Greek letters. I'm visiting Emma's southern, Greek-dominated university during homecoming week, and ever since I arrived, I've had to step over fraternity brothers and sorority sisters sprawled on the floor of the student center, building floats for the upcoming parade.[1]

Even in sweats, Emma is stunning. Her eyes are tired, and she might be nursing a hangover, but Emma is effortlessly beautiful, the kind of girl who surely sparks envy among her peers. What's more, Emma is smart and very invested in her studies. A junior honors student, Emma has a double major in finance and psychology to go with her status as an officer at Alpha Alpha, her university's most prestigious sorority. If there is a social hierarchy at this school—and there pretty much always is—then Emma is at the top of it.

On the day we meet, however, she is *so over* being a sorority girl. Emma is incredibly unhappy, and she's not afraid to show it. She's frustrated with Greek life and especially with the pressures that come with being so high up on the food chain. It makes life ridiculous, Emma says. And social media just makes things worse.

Emma rolls her eyes so many times when talking about sorority life that I lose count. "Every house has its own reputation, and its own facet of people that fit into that house," she tells me. "I'm not exactly congruent with the type of people that are in my house, so that's been a challenge in getting to know people and, you know, where I fit in. I don't really like to party or drink, and my sorority is known for that." She pauses to roll her eyes. "We're also known for working out excessively, not really eating—that's pretty much every sorority on campus." Emma's tone is sing-song, as if she thinks everything about sorority life is absurd. Her facial expressions are exaggerated, almost theatrical. "[My sorority sisters] don't like to be challenged in relationships and questioned because that would require having an opinion and a brain." On the surface Emma might sound snobby, and she is certainly angry, but I detect an unmistakable sadness beneath everything she's saying. Her comments prompt me to ask why Emma belongs to a sorority at all. "If you're not in a Greek organization, it's extremely hard to make friends," she says. Then she grins a bit wickedly. "If you *are* in a Greek organization, it's extremely hard to make friends, but people automatically *have* to like you if you're wearing your letters. Or, at least, they have to automatically be nice to you because there are repercussions if they're not. Not because they're good people and they like you—it's that, you know, they'll get in trouble if they don't. So it forces a sense of camaraderie and this illusion of belonging on the part of the individual. It just forces the system to expedite the whole making friends process."[2]

Despite Emma's problems with Greek life in general and her prestigious sorority in particular, she claims she's happy she joined because "I would rather the social stigma of [Alpha Alpha] be on me than the social stigma of someone who's *not* in a sorority. It's the lesser of two evils."

Particularly hellish, I learn, is the period when sororities on campus are courting new members. Emma and her sisters spend up to twelve hours daily, "practicing chants and going over matching and how to bump someone, and just all the logistical nonsense that goes into formal recruitment on the side of the chapter." This year, Emma is trying to "limit the hypocrisy" on her part (eye-roll), so she got herself excused from recruiting. She didn't want to stand there and tell new students how her sorority "is everything."

After a couple years in Greek life, Emma has very little faith in a person's ability to be honest about who they are. "I think, in general, people are not very authentic," she says. "I'm probably jaded, and not maliciously, I just think there's so much pressure to fit in with the mold, that there's no way there's this many people that actually behave that way, and if there are, then God has a sick sense of humor, and we're all clones of each other."

This is when she turns to social media.

"People do [social media] for the 'likes.' People take pictures, experience things, go places for the reaction that they're going to get on social media." Emma hates the fact that people she knows on campus will do things just so they can post pictures on Instagram. "Obviously, I don't go around saying, 'I think all of you are fake and snobbish and unintelligent' and things like that," Emma says, even though that is obviously what she believes. "That is my authentic opinion, but I would be a leper if I shared that opinion. So I definitely am guilty of just going with the flow, wearing my [sorority] letters, because of the reputation that I have because of it, because of the esteem that my house holds on this campus. . . . Why make it more difficult for yourself if you have the opportunity to make it easier by wearing a shirt, or by wearing a button, or by telling someone you're an [Alpha Alpha], or by telling someone you're part of this or that? I don't think anyone should make it harder on themselves unnecessarily. So I am hypocritical in that sense, that I have all these opinions, but I am very much one way with the people that I am close to in my family and another way with the general public."

Emma is clearly disgusted by the state of things and by how social media exacerbates inauthenticity among her friends and peers, but she is also disgusted with herself for going along with this charade. Emma has both Instagram and Facebook accounts. I ask Emma if it's tiring, being one way in public but feeling like a completely different person in private.

"Yeah, it is, but I think that everyone is like that," she says. "I put things online that people are going to respond well to. I stay far away from any political platform, religious platform. I don't post statuses on Facebook. I think that my life is interesting to me, and I enjoy it. My mother loves hearing about it, my boyfriend loves hearing about it, but outside of ten or fifteen people, my fourteen hundred Facebook friends don't care. Sometimes I'll upload pictures to Facebook and Instagram if I think I look pretty and I'm going to get a lot of 'likes,'" Emma admits. Even as she dismisses her Facebook friends, she can't help but mention the impressive number of them. Emma participates in social media, but she thinks it has spawned a "very weird culture." "People used to do things and *then* post them, and the approval you gained from whatever you were putting out there was a byproduct of the actual activity," she says. "Now the *anticipated* approval is what's driving the behavior or the activity, so there's just sort of been this reversal. Not everyone is that way, obviously. I just feel like the majority of this campus [is], judging by some of the things that I've overheard in my sorority house."

Being a sorority sister exacerbates this, Emma thinks. "If someone who was highly associated with Alpha Alpha were to do something like protest for gay rights and put it on social media, people would be taken aback," she tells me. Politics—anything potentially controversial—is off limits for sorority sisters. "If someone who was *not* in a Greek organiza-tion were to protest for gay rights and put it on Instagram or Facebook, that would make more sense. It would be more congruent with the cat-egory we lump those people into." Emma rolls her eyes. "I just said 'those people,' so there you go. That's how it is. It's very Greek/not Greek. . . . Even amongst Greek life, we have tiers. We have top-tier sororities, we have secondary sororities, we have third tier." Emma goes on to name the sororities and fraternities on campus and how they associate, or don't, with each other, and how anyone on campus who isn't a part of Greek life is called a GDI (God Damn Independent; this was the first time I hear this acronym, but not the last). The careful social hierarchy the Greeks have constructed plays out online in a big way, Emma says.

But not necessarily in the way you might think, with a constant stream of wild party photos. Far from it. The era of posting pictures of frat boys doing keg stands and girls chugging beers at parties is over. Emma tells me, "We have social media workshops about what is appropriate for social media because my sorority engages in some pretty questionable behavior that needs to be discussed, that we can't put online." Everyone knows that bad behavior happens; you just need to make sure not to take pictures of it to broadcast on Instagram. Emma thinks that her sorority does "more bad things than good things," which is also why the sorority helps its sisters choose which photos to put up on social media. She shakes her head. "The hell that would come down on you if you were to share or be tagged in any kind of photo that is congruent with our actual behavior and not this '[Alpha Alpha's] sisters are smart,' and '[Alpha Alpha's] sisters are responsible,' and blah, blah, blah. You would be sent to standards, and you would be ripped a new one."

Emma's sorority (like many sororities and fraternities, she indicates) has a designated person who monitors all social media activity among the members. This person at Alpha Alpha has pseudonymous accounts on various social media platforms, and if she "likes" any posts or photos a sister puts online, this means they are inappropriate and must be taken down immediately. "Your picture better be off every social media platform" within twenty-four hours, Emma says. Emma's sorority also monitors the social media accounts of those seeking to join. The sisters take screen shots of behavior that is not up to Alpha Alpha standards. "Any behavior that is not congruent with being an [Alpha Alpha] is questioned," she says.

Despite being annoyed by her sorority and overwhelmed by the expectation to be a public, personal example of the kind of girl the sorority wants to advertise to the world, Emma finds herself doing exactly that—projecting a fake version of herself. "It's just interesting," she tells me. "I put forth [Alpha Alpha's] version of myself on social media." Everybody else is doing it, so she goes along and does it, too. In part, Emma is driven by what she sees everyone else posting. She doesn't want to appear to be falling behind, having less fun, being less spectacular. Even a beautiful, accomplished young woman from the most prestigious sorority on campus sometimes sees things on social media and starts to feel bad about herself.

As Emma's reflections progress, cracks appear in the hardened attitude she's trying to project. "People share the best version of themselves, and we compare that to the worst version of ourselves," she comments. "I know I've done it." "Over the summer, everyone was in bathing suits all the time, posting fun pictures," she went on, "and I said, you know, 'Wow, I want that.' Like, I forget what I have. Like, I would like to look like that."

Here, Emma hits on something that many students expressed, though none quite so succinctly or vividly: that on social media people are seeing only the "best versions" of their friends and acquaintances, which have been edited and curated down to the last glorious detail. Yet, behind the screen, they are still themselves, with all their imperfections, insecurities, and perceived failures. To hear Emma say this is startling and telling because she seems to be the epitome of what others would envy and wish to be. But in the end, Emma is just like everyone else. Social media can bring down even the most popular and successful students on campus.

Remaining rational while viewing one's Facebook feed and other people's Instagram photos is no easy feat, according to Emma. She knows on an intellectual level that people are posting only happy things and pretty pictures and not sharing anything difficult in their lives. She knows that not everyone is as perfect and as happy as they seem on their social media accounts. But being exposed nearly constantly to everyone else's veneer of happiness can get to a person after a while. Emma may know that what she's seeing isn't reality, but it's one thing to know this and another to scroll through an endless stream of beautiful, smiling faces while sitting alone in your room.

"I think I'm a very confident person, and I recognize that the pictures I'm seeing on social media are not anyone's struggles or anything like that," Emma says. "I know not to compare myself, you know, when I'm having a bad day, to these images that I see because that's not *them* having a bad day. So I think that it can very much affect self-esteem if you let it, and if you are under the illusion that, you know, this person is perfect and this is how they are all the time, nothing goes wrong in their life, blah blah blah." Emma offers the example of a girl she knows who is on antidepressants but who posts pictures "throwing her hands up and jumping in the air, so happy." Emma realizes "there is a lack of congruency between who we are as people and the image that we put out there." People shouldn't put a lot of stock in what others share on social media, she says, because "it's very shallow," but sometimes you just can't help it. "I feel like, [on social media], it's the best version of people," Emma comments for the second time. "It's just sort of a slap in the face. It allows a platform to brag, and if you are experiencing, you know, [lonely] feelings, it makes it a little harder, a little more obvious that you're alone or lonely, or whatever the case may be."

Emma feels immense pressure to live up to the image of perfection that she sees on social media—not to mention living up to the expectations of her high-achieving, image-conscious sorority sisters. But there is also the pressure to be perfect for an entirely different audience: future employers. "We have all these talks [on campus] about employers checking social media regularly," Emma says. "There's a lot of opinions that are formed based on people's social media platforms, rightly or wrongly, so just the gravity that it has on society today is a little unnerving. People cannot get jobs, or not get interviews because of what they put on social media." Students have this drilled into them, and it makes them incredibly anxious about what they post. Here, too, is pressure to be some perfect version of yourself.

The only time people are really honest about anything on social media, in Emma's opinion, is when they are anonymous. Protecting one's identity is paramount. You need to go to sites like Yik Yak—a popular app at colleges, a sort of anonymous Twitter feed tied to a person's geographical location—to see anything "real" about what's happening on campus. "I would like to pretend that I'm a good enough person not to get on Yik Yak and see what is being put out there," Emma tells me. Yet she's not—she's on her university's Yik Yak, just like just everyone else. "But it's Animal House. . . . If I were looking at this university as a senior in high school again, and I had access to the profiles of people in Greek organizations at [my school], and Yik Yak was a thing, I would not be here."

Unlike Facebook and Instagram, where what you see is a carefully crafted showcase of a person's best, happiest, prettiest moments, what you see on sites like Yik Yak is reality, according to Emma. Nobody is honest on platforms that require real names to be attached to their posts, but give a student body anonymity, and people go to town—often in the ugliest of ways. "I feel like unfortunately [what's on Yik Yak] is a pretty accurate representation of what happens at [my school]," Emma explains. "It is calling people out for everything. People use specific names, like 'Sarah is a whore. She hooked up with this person, this person, this person, this person, this person, this person.' And, 'So and so does cocaine.' And, '[Delta Betas] are,' and any kind of degrading comment you can think of."

The people who get attacked, maligned, and degraded on anonymous campus feeds are usually women, in part, Emma believes, because men are more drawn to Yik Yak. "Girls don't really post on Yik Yak," she says. "I'd like to think it's because we know how it feels to be spoken of poorly, in such a public arena, and so we wouldn't want to put anyone else through that." But maybe it's the women Emma knows in her sorority who aren't posting, since her sorority's sisters aren't allowed to post on it—ever. "The week that [Yik Yak] got popular, we were called into [Alpha Alpha]. '[Alpha Alpha] does not comment. [Alpha Alphas] do no post on Yik Yak,'" Emma mimics. Underlying this edict is the belief that the anonymity promised by the app is an illusion, and Emma's sorority is not willing to take any chances when it comes to its very esteemed, yet very precarious, reputation. "It's very much the whole Snapchat thing; it doesn't disappear, there's a record of it. Yes, after twenty-four hours, [Yaks are] no longer at the bottom of the page, but the IP address that's associated with your phone is associated with everything that you do through the app, and that's dangerous. [Alpha Alpha] is concerned about what we put out there. We have been on social probation in the past for doing things, like getting caught doing cocaine in the house, so we're very careful about what we're putting out there, regardless of the alleged anonymity that Yik Yak provides." The fact that inappropriate posts are seen as the equivalent of using cocaine tells you how concerned people are about social media.

Despite Emma's frustrations about what she sees as the inauthenticity of social media, the drama it causes, and the ways in which it makes her feel bad about herself, she doesn't really think about quitting any of her accounts—though I eventually learn that she has tried to do so. Sort of. She once gave up social media for Lent but has no plans of doing something like that again. For Emma, there's "entertainment value" in social media, and she doesn't want to miss out on it. She deserves the entertainment, really, given all the stress social media causes her. The same goes for Emma's relationship to her smartphone. She simply can't be without it.

"If I don't have my phone, I feel empty," she says. "You know, if it's not in my hands, I feel like I'm forgetting something, I'm missing something. Even though when I'm with people, I make a conscious effort not to be on my phone at dinner, at lunch, in social situations. . . . I will leave it in my bag at dinner. I won't take it out at dinner with my boyfriend or with friends or anything like that. Even if we're eating dinner at home, it's not on the table in front of me. But I've never, I *don't* leave it at home, because if something happens, I need to have it." Indeed, Emma considers it dangerous to be without it. At this point she reflects on how there has been a number of sexual assaults reported on campus recently, and because of this she "would hate to be without her phone." Today, especially if you are a young woman, a smartphone "is very much a necessity."

Although Emma is an extraordinary young woman, when it comes to her struggles with social media, she is absolutely ordinary. She is concerned about getting "likes." She gets caught up in the comparison trap, constantly seeing the "best versions" of people on social media, which makes her feel bad about herself. She believes that keeping up online and being always available on her smartphone is almost a job (though one that comes with definite entertainment value). But Emma is typical with regard to the pressures she feels to maintain a positive, nearly perfect appearance on her social media accounts, and she is extremely careful about what she posts as a result.

I ask Emma if there's anything else she wants to mention before we end our interview. "I think the weight of [social media] is a little concerning," she says. "But we do it to ourselves."

THE HAPPINESS EFFECT

During the course of the last decade, I've traveled to well over a hundred colleges and universities to discuss my research about college students, sex, and faith. In doing so, I've had the opportunity to listen to students all over the United States describe their concerns and struggles with life on campus. We've talked over dinner and in small groups for coffee and in classes. Inevitably, these discussions widened to involve questions about meaning and purpose in general, about identity, and about what it's like to be a young adult in the first generation that has grown up with social media.

It doesn't seem to matter whether I am visiting a Catholic university in the Midwest or a private-secular college in the South; social media is on everybody's minds of late. Students can't stop talking about it. Questions about the various stresses it provokes in today's college experience are nearly constant. Students discuss the notion of the "real me" versus the "online me" and the dissonance they feel between these, the pressure to document publicly a certain kind of college experience, their fears about making themselves vulnerable on social media, and their worries about how to maintain real, meaningful relationships when a seemingly artificial online world dominates their social lives. Students want to know what their peers think about social media and whether they experience the same struggles. They want, in other words, information about how their generation is handling one of the most significant and dramatic cultural shifts of our time. Most of all, they want to know that they are not alone in feeling the way they do.

That students are aware they are splitting themselves in two—that they somehow *have* to do this to operate effectively and safely online—has been particularly fascinating and worrisome to me. It's not as though students don't talk about the joys of social media—they do. They love the ease of the connections it offers among far-off family and friends, and many of them love the ways that social media affords a certain creativity and opportunity for self-expression. But they are also exhausted by it. There are many things with which they struggle, things that unnerve them, that make life difficult and even painful, and they don't know where to turn to talk about them with any honesty. Many young adults experience some kind of alienation because of social media, but they are further alienated because they don't see a thriving public discussion about the struggles they are experiencing—perhaps because those struggles aren't as racy or extreme as the ones that are the stuff of newspaper and magazine headlines.[3]

Media coverage of social media often focuses only on the belief that this generation is the most narcissistic generation ever, or on the scariest examples of what happens to young people online—predatory behavior, risqué pictures that get circulated, and cyberbullying and related suicides.[4] Clearly, these are important issues. But while young adults and college students buckling under the pressure to project a false self online may not be as sexy (literally) as teens sexting nude photos to each other, it's a pervasive struggle, and we need to talk about it with them.[5]

So, beginning in the spring semester of 2014 and continuing on through the spring semester of 2015, I visited thirteen colleges and universities to conduct private, one-on-one interviews with nearly two hundred randomly sampled young men and women. I also conducted an online survey of students who volunteered to share their opinions in a series of essays.[6] After both the interviews and the online surveys, I feel confident in saying that the social media world is a far less scary place overall than the press would have us believe, and that the young adults with whom I spoke are as smart and thoughtful as ever. They are doing their best to navigate a dimension of culture so new and different—and so pervasive— that it sets their generation apart. Like most of us, young adults have no choice but to confront social media in their lives.[7] It shapes their identities, their relationships, and the ways in which they make meaning—or don't. It troubles them, but it's also a sphere in which they are learning to work out the dimensions of their social lives and identities in much the same way that my generation did as we rode our bikes through the neighborhood and hung out on the playground.

It didn't take long to find out that one of the most central concerns college students struggle with, however, is the feeling that they are constantly monitored on social media—potentially by anyone and everyone. When they were still in high school, they were wary of their parents, their teachers, and the admissions officers at the colleges they hoped to attend. When they arrive on campus, they believe that future employers will be assessing their every post. Emma had the added pressure of her sorority, which was literally monitoring everything she did online. I heard similar accounts from Greeks at other universities I visited, but everywhere I went students talked about people who were monitoring them: athletic coaches, Student Affairs staff, professors, the Career Center, Campus Ministry. One student who had dreams of holding political office was concerned about future constituents. Students even worried about other students who liked to keep an eye out for potentially offending or negative posts by their peers.

The result is that students create carefully crafted, fantasy versions of themselves online. But on platforms that allow for anonymous posts, things get really dark.

In the best of circumstances, apps that come with the promise of anonymity and impermanence—like Yik Yak, Snapchat, and the anonymous Twitter feeds and Facebook groups students create for venting, confessing, and other types of honesty not found elsewhere—serve as cathartic forums in which highly pressured and highly monitored young adults can finally be themselves. Sometimes they are playful and silly. Yet the kind of commentary that often bubbles up can be incredibly vicious, revealing a nasty underbelly to the student body that shocks even the students themselves. Simply put, anonymous forums tend to degenerate into cesspools of obnoxious, cruel, and sexist comments, in which students treat each other (and their professors) in the worst possible ways, and entire campuses find out exactly how vile and racist certain members of the student body can be.[8]

A "work hard, play hard" mentality often prevails on campus. Extremely stressed, high-achieving, incredibly busy college students work extraordinarily hard at their studies, sports, and activities during the week but then party like crazy and drink as heavily as they can on the weekend, believing they "deserve" to engage in such behavior because they are so overburdened the rest of the time. This mentality seems to transfer online. Students feel they must maintain a perfect, happy veneer on Facebook and other profiles attached to their names. They must be that high-achieving, do-no-wrong, unstoppable, successful young woman or man with whom everyone would be proud to associate, to have as a son or daughter, to boast about as a resident assistant or a member of a team, and, eventually, to hire. Many students have begun to see what they post (on Facebook, especially) as a chore—a homework assignment to build a happy facade—and even to resent such work. Then they "play hard" on sites like Yik Yak where they have learned to unleash, to let go, and often go a bit crazy—even if people get hurt in the process. They deserve to let loose, after all, since it's tiring to be so perfect all of the time.

The colleges and universities I visited were incredibly diverse—geographically, ethnically, socioeconomically, and in terms of their religious affiliation or lack of one and their level of prestige. Yet across them all, one unifying and central theme emerged as the most pressing social media issue students face:

The importance of *appearing* happy.

And not just happy but, as a number of students informed me, blissful, enraptured, even inspiring. I heard this at one of the most elite private institutions in the United States and at a school that doesn't even appear high enough on the rankings for people to care where it ranks. The imperative transcends every demographic category.

The pressure to appear happy seems universal. In fact, it came up so often in the interviews that I asked about it directly in the online survey. Students responded to the following statement by indicating "yes" or "no" (or "not applicable"):

I try always to appear positive/happy with anything attached to my real name.

Of the students who chose to answer this question, 73 percent said yes. Only 20 percent said no.[9]

What makes these data especially important is how ubiquitous social media has become.[10] Out of the 884 students who participated in the online survey, only 30—a mere 3 percent—said they did not have any social media accounts. For the vast majority who do, students were asked how often they check their various social media accounts per day. The results are shown in Figure I.1.

It's clear from these data that students are checking their accounts compulsively throughout the day, with approximately 31 percent of students checking at minimum twenty-five times and potentially up to a hundred.

Given the amount of time young people spend on social media, the pressure to appear happy online can become overwhelming. Adolescents learn early how important it is to everyone around them that they polish their online profiles to promote their accomplishments, popularity, and general well-being. They practice this nearly constantly in their online lives and this has a tremendous effect on them—emotionally, in their relationships, and in their behavior on social media. For better or worse, students are becoming masters of appearing happy, at significant cost. This is what

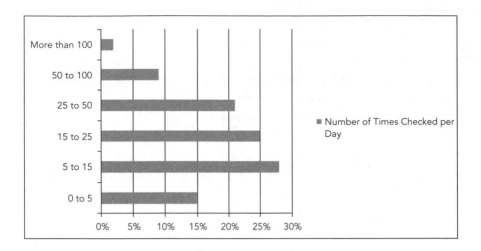

Figure I.1 | Daily Levels of Social Media Usage

I've come to think of as the "happiness effect." Simply put, because young people feel so pressured to post happy things on social media, most of what everyone sees on social media from their peers are happy things; as a result, they often feel inferior because they aren't actually happy all the time. The chapters that follow all explore themes that emerge from this larger issue.

The happiness effect not only has implications for the emotional health and well-being of the young adults in our lives but also has repercussions for the way we parent and how institutions of higher education help (or hinder or ignore) how students experience college and interact with the wider world. Anyone who works in higher education as faculty, staff, or administration, who is a parent, or who works as a high school teacher or counselor needs to think broadly and deeply about what we are doing when we teach our children that proper online behavior requires the appearance of happiness—because they are learning it *from us*, even if we do not practice it ourselves—and boy do they notice when we don't. We need to reflect, very seriously, on the message we are sending. It may seem like a logical and responsible thing to teach our young people—post only happy, flattering, achievement-oriented things—but when the young people receiving this message are on social media nearly constantly, what seems like a simple rule can have extraordinary consequences.

If students have not already imbibed the lesson that a happy appearance is the right appearance on social media, they will certainly learn it once they get to campus, whether from peers, professors, or the career center. The notion that one must carefully craft, cultivate, and curate (what I've come to think of as the three Cs of social media) public profiles at all times permeates the lives of students. It's a given. This doesn't mean that people don't rebel against this expectation and find clever ways around it, but a young woman's or young man's ability—or inability—to live up to this pressure and constant expectation infects every dimension of his or her online experience. It influences a person's vulnerability to bullying, people's biases about gender and race, whether students express their religiosity and politics, whether they choose to quit social media for a time or permanently, students' relationships with their smartphones, and their sense of place (or lack of one) in their larger social worlds. It influences what students post and how they post it, whether they post at all, their sense of their future, how they feel about the "real them" and how they find authenticity (or don't) in their relationships with others, the ways in which they compare themselves to their peers, and whether or not they feel accepted or isolated socially.

The pressure to appear happy can also warp how students see themselves—as successes or failures—and whether they have an enjoyable or discouraging college experience, or even one that they feel is inferior to what they see online in the experiences of others. Further, the energy required to maintain such appearances on social media can be exhausting. It forces many students to hide who they really believe they are and teaches them that anything that doesn't present "a happy face" is best kept out of view. It also teaches them that provocative opinions do not belong in the public sphere—provocative opinions get you rejected by friends and acquaintances, and perhaps even by the employer of your dreams. Students have learned that signs of sadness or vulnerability are often greeted with silence, rejection, or, worst of all, bullying. The importance of appearing happy on social media—the *duty* to appear happy—even if you are severely depressed and lonely is so paramount that nearly everyone I spoke to mentioned it at some point.

And some students spoke of virtually nothing else.

I

IS EVERYBODY HANGING OUT WITHOUT ME?

COMPARING OURSELVES TO OTHERS AND THE IMPORTANCE OF BEING "LIKED"

[Social media] gives this false image that you're living a perfect life, that your life is just like a fairy tale. Like, everything is good at all times because you don't want people to see you at your low times. You want them to see only the good times so that they go, "Wow, I want to live like him."

Michael, sophomore, public university

I mean, it's not necessarily like, "My Instagram is better than your Instagram," but they want to one-up people to show, like, "I'm in a better place than you are and I want to prove that to you and show you that I'm extremely happy," kind of thing. Like, "I'm doing better. I'm in a better place, and I'm much happier than you are."

Laura, junior, private-secular university

MARGARET: FACEBOOK CAN REALLY GET YOU DOWN

Margaret has soft, dark curls, glasses, and, from the moment she walks into the room, a nervous demeanor. One of the first things I learn about Margaret is that, though she is only a junior, she is already engaged. She proudly shows me her ring. Both Margaret and her fiancé are on the ballroom dance team at their Midwestern Christian college. Margaret is involved in theater and choir too, and she volunteers at a nursing home—she wants to be a social worker. Ballroom is really the center of Margaret's life, though, and most of her girlfriends are involved in some way with it, or at least are highly supportive of her participation on the team.

On one level, Margaret seems very self-possessed. She's excited to get married, loves her activities, and is already sure about her choice of careers. But Margaret's mannerisms—she sighs and sighs often, great long exhales, her voice shakes, she hems and she haws and backtracks and then stumbles forward again as she talks—reveal a young woman who is unbelievably insecure and stressed out by much of what life throws her way.

Especially when it has to do with social media.

I don't even have to ask Margaret any questions about social media. She launches into the topic on her own, in particular how she's really been trying to stay off Facebook lately. It's been "a struggle" for her. "I'm not as involved on Facebook because I compare myself a lot to other people, and Facebook is a really easy way to do that," she says. "You can just click on other people's posts, see everything that everyone's doing, and when I see that on Facebook, I think, 'Oh, they're doing all that, they're just so happy,' because no one puts anything bad on Facebook. They only put the greatest things going on in their life, so their life is only a *part* of their life that they're showing that's good, but everyone still has the bad." Even though Margaret knows on an intellectual level that people who show only happiness are posting selectively about their lives, she still has trouble preventing it from upsetting her. "I forget about all the great things that are happening in my life, that I need to notice," she says. "It's just really been an easy trap for me to get into, just comparing myself, and Facebook does not help."

The more Margaret and I talk about social media, the more pronounced her mannerisms become. In addition to sighing and speaking in a shaking voice, Margaret pauses occasionally to cover her eyes with her hands or to lay her head down on the table. She tells me several times that she's trying hard to stay off Facebook, that she only posts a status maybe once every three months—usually it has to do with ballroom dance, or a major event, such as when her cousin got engaged—though she still accepts any and all friend requests. If she does go on Facebook, she gets on and off quickly, but these brief visits upset her. Margaret gets distracted, she says, she stops "living in the moment," and her "mind is everywhere." There was a time when Margaret was on Facebook constantly, but she soon found out that it "wasn't good for [her]." "It wasn't helping my self-esteem, so I just decided to stop," she says. "[Because of Facebook,] I still didn't recognize what I had going on in my life that was important."

Margaret has a slow, antiquated smartphone, and although she resents this, in many ways it has been a blessing. She can really only go on Facebook if she's at her laptop, and that has helped her to stay away. Margaret loathes what happens to her when she's on Facebook. "If I spend a large amount of time on Facebook, there are things that I want to happen on that Facebook page," Margaret explains. "I want people to be IMing me, and sometimes I do not take the time to start IMing them first but I just want them to see, 'Oh, I'm on Facebook. Please, like me.' Like, 'Start that conversation.' Or, 'Make sure you "like" my pictures or "like" my post.' And then if you don't 'like' it enough, then I grade myself that I'm not as good and that's stupid," Margaret adds with a long, frustrated sigh. Margaret would like it if people communicated more by email, but students at her college tend to use Facebook instead, and they are constantly checking it. "I don't want to be a slave to Facebook," she tells me. "I think it's just not worth it."

Margaret realizes that social media isn't going away, though. She keeps trying different platforms and then getting disillusioned soon after setting up a profile. Margaret thinks they all "suck in time." Social media is for people who can multitask, and Margaret is not one of them. She mentions repeatedly the hours and hours that a person can waste on social media—she marvels at this, is maddened by it, and wishes she had the willpower to tear herself away.

When Margaret *was* on Facebook all the time, she became obsessed with "likes." "I would just basically compare the number of 'likes' that other people got to the number of 'likes' I got, and then I wouldn't be happy with those and then, once again, I would grade myself based on that," she says. Not getting enough "likes" made Margaret feel she wasn't popular enough, or even that she wasn't friends with the right people.

That Margaret talks of "grading" herself based on numbers of "likes"—that she actually uses the word "grade"—is striking, and telling, too. As with "likes," grades are quantifiable, and designed to tell a person if they are doing well or perhaps, failing. In *The Culture of Connectivity: A Critical History of Social Media*, José Van Dijck names what she calls the "popularity principle," the idea that, on social media, quantity equals value. For example, the quantity of Facebook "friends" one has can affect one's sense of personal social value. The same goes for the way people like Margaret respond to that "like" button. "Online quantification indiscriminately accumulates acclamation and applause, and, by implication, deprecation and disapproval," Van Dijck writes. "Popularity. . . . thus not only becomes quantifiable but also manipulable: boosting popularity rankings is an important mechanism built into these buttons."[1] For someone like Margaret, sitting there and watching the number of "likes" go up on her post—or not go up at all—is akin to watching her popularity rise or fade away. So potent is the experience of getting and not getting "likes," that Margaret givies herself a "grade" in relation to them. Through "likes," Margaret has learned to quantify her self-worth—or lack of it—and too often Margaret has found herself getting failing grades in the category of popularity.

Because of what she saw on Facebook, Margaret used to spend all of her time thinking she should be "somewhere else," and that prevented her from being happy with where she actually was. "Self-esteem for me has always been a battle," Margaret explains, sounding more and more fatigued. "You can do a lot of damage on Facebook. Like, I was never bullied on Facebook or anything but, to be honest, I've always sort of bullied myself. I put myself down on a lot of things and judge myself really critically."

As with Margaret giving herself "grades," this comment—that she "bullies herself"—is both stunning and telling. The more sensitive a person, the more emotionally vulnerable, the worse he or she fares on social media. The students I interviewed who suffer from insecurity, who have anxiety about their social standing, who fret about how they are seen by others, are the ones who are drowning on social media. These are the young men and women for whom social media is a highly destructive force, and they stand out from the many other students I met who feel ambivalent about social media and the rare few who really thrive on it. Margaret is the kind of young woman who should make us concerned—her real-life vulnerability is magnified by social media.

Margaret actually spoke of worrying about other, younger kids who go on Facebook and "don't feel accepted," believing they could become "suicidal, just based on what they're seeing." I ask Margaret if she ever felt this way when she was younger, which provokes one of her deepest sighs yet. "Ahhhhhhhh. Possibly," she says. "I definitely have had times when I was in high school that I was very depressed." Margaret knows that personal suffering is not to be shared or discussed on social media, however, so she kept this to herself.

"I think that not doing as much social media has helped me, for sure," she says toward the end of our interview. "Yeah, and the problem is, this world is moving forward, and I don't really care to go," she adds with a sad laugh.

MICHAEL: EVERYTHING YOU SEE IS FAKE

At a public university in the Southeast, I meet Michael, a classic frat boy in looks and mannerisms who *hates* social media. He was a heavy user for years but felt so lonely and bad about himself all the time that he forced himself to quit. Michael's Facebook page still exists, but he hasn't been on it in ages—his last post was four years ago, he tells me. Now, his only contact is really through his fraternity's Facebook page (which monitors the brothers online much like Emma's sorority), and life is better as a result. Before I learn all of this, Michael tells me his mantra for himself, which turns out to reveal a lot about who he has become since he stopped comparing himself to others online.

"Just be you," he says. "Just don't be ashamed. Who cares what other people think? Don't try to be something that you're not to try to impress someone else, or for whatever reason that you may do it. Just be *you*."

For Michael, social media is all about performance—a performance of your "self" in an effort to impress other people. Michael is cynical about social media and doubts that there are many people who are really on it for themselves. "I feel like social media is very *not* authentic," he says. "It's kind of an ego thing, I think. It's all about the 'likes.' It's all about, 'What can I do to show everybody else how great my life is?' I just don't feel the need to put my information out there."

At least not anymore. But high school was another story.

"I would post daily statuses about just anything. You know, 'I'm doing this right now' or, 'I'm eating breakfast right now.' It just got old to me"—Michael pauses here for a long moment—"and it was redundant and I felt like it wasn't a necessary thing for me to be doing, so I stopped."

A rather unlikely source prompted Michael's first step back from social media—his grandmother. When she got a profile and friended him, she went onto his page and was appalled by how much he posted, the kinds of updates he made, and the fact that he used so much profanity. This was not the grandson she knew, and she told him so. Michael realized she was right, and once he began thinking this way, he felt "there was no point" to posting anymore. It has been a huge relief. He no longer has to worry about posting, getting "likes," regretting something he put up after the fact, or maintaining an online image. But the biggest relief is not getting pulled into the game of everybody comparing themselves to everybody else.

"[Social media] is becoming more and more how can we show other people how happy we are instead of just, you know, spending the time with the other person," Michael says. "The best example I can think of is last year, for one of my friend's birthdays, we went to [a baseball] game and it was a big group. It was eight girls and eight guys. We got down there and we made this whole big thing. It was an hour long where we had to take pictures together so that we could post them. We had to take them by the water. We probably took two hundred pictures. It was ridiculous. And, you know, of course we go to the game. We come back and they post all the pictures, and from the pictures it looks like everybody's smiling and everybody's having a great time, but while we were taking those pictures, nobody was having a good time." Michael laughs here and shakes his head disbelievingly. He goes on to talk about how the girls were all worried about their hair, and everyone was worried about which photos had the best lighting and whether their smiles were good: "It's a stressful thing because they want to get the best picture so that they can post and show everybody. I think that's what [social media] is turning into."

And Michael is glad he's exempt.

"It's comparing yourself to other people, I think," Michael tells me, trying to explain what's behind the need to take two hundred photographs and then worry about choosing only the best ones to post. "If one person posts a selfie and gets ten 'likes' and you post a selfie and get eleven 'likes,' you may feel better than that person. You may feel like you're superior to that person. It's all about our mindset. It's all about where you place your values. Where do you get your confidence from? What do you think matters?"

Social media sells Michael's generation on the notion that what matters most of all is what everybody else is doing—and managing to fit yourself into that mold of what other people idealize as opposed to who you really are. People's confidence comes from this misguided place, Michael thinks. People see themselves as superior when they get more "likes." The worst part, Michael believes, is that what you see and what people post on social media is all fake. It's a carefully cultivated facade.

"[Social media] gives this false image that you're living a perfect life," he explains. "That your life is just like a fairy tale. Like, everything is good at all times because you don't want people to see you at your low times. You want them to see only the good times so that they go, 'Wow, I want to live like *him*.'"

Michael doesn't think that his generation is more self-centered or self-obsessed than any other—it's just that his generation happens to have the tools "to show everybody" how self-centered and self-obsessed they are. "Everybody wants to be noticed," he says. "Everybody likes feeling approval. They all like it when other people like them." Social media just gives people a way to amplify this—a really big, public way. Michael mourns this reality. It saddens him that he doesn't know what it's like to live without social media, since it's been around as long as he can remember. He believes it "affects people's happiness for the worse." When people get that coveted public affirmation they are so desperately seeking—when a post is "successful"—"that might make you feel good in the moment," he says. "But then you scroll down your Facebook feed and you see someone else who posted their pictures, like, one of the two hundred pictures they took that day, and it looks like they're having a great time, then you ask yourself, 'Why am I not? Why can't I have as much fun as they are?' *Even though* they were in the same *exact* position, they weren't having a good time while they were taking that picture. It just gives off the impression that they were, that their life is perfect."

Social media can make us feel, perhaps ironically, *really* isolated.

"You see all these other people that you're friends with, and they're doing all of these things," Michael says. "Only, you see the good things they post on Facebook or Instagram or Twitter, and it just makes you feel like your life isn't as meaningful as theirs is. And that can be a pretty lonely feeling."

In addition to Michael pointing out, like so many of his peers, that people only post about "the good things" on social media, his talk of the loneliness, the isolation, and even the feeling of meaninglessness this can provoke was also common commentary during my interviews. Students talked of this strange, vicious cycle that being on social media regularly draws them into—the constant going online to check out what everyone else is doing and what your friends are up to, only to find that this very form of "socializing" ends up making them feel all the lonelier, even though—at least in theory—connectivity is one of the central ideas (if not *the* central idea) behind social media.[2]

Michael has such strong opinions on all of this that I ask if he also felt this way when he was an avid Facebook user. "I *did*," he responds. "I think that stuff has gone away a lot, now that I've been off it. It's just, you know, highs and lows. I personally don't really like it."

LAURA, MATTHEW, AND HANNAH: SEEING EVERYBODY'S HAPPINESS IS JUST PART OF THE DEAL

Not all young people are as conflicted as Margaret and Michael. Plenty just accept that comparing oneself to others is a fact of social media. Having a window into other people's lives is what you sign up for, so either deal with it or quit (if you can manage to).

Laura, for example, a junior at a private northeastern university, is very conscious of how everyone compares themselves to others on social media, but she seems like a detached observer—she's interested in (and a little disgusted by) the phenomenon but doesn't let it get to her much.

But there are many things that bother Laura about how social media is changing our lives: the fact that everyone is constantly on their phones, how everyone is obsessed with taking pictures of everything they do so they can post them and prove what they've done. "With social media you just never get to live in the moment," Laura says. "You remember things by the picture that you took, not by the memories that you actually made there."

Why is everyone so obsessed with documenting everything they do? Because there is a competition going on. "I feel like some people are more worried about, 'What are other people doing compared to what I'm doing right now? Am I doing something better than somebody else? Oh, yeah. I'm awesome!' Or, 'Is somebody else doing something better than I am right now, so I should be doing something better than what I'm doing right now?' It's one-upping people," Laura goes on. "People try and one-up each other all the time." Laura says that the only thing her friends do these days is "sit around and drink and take pictures of each other and post it on Instagram." Then once the picture is up, they compare who got more "likes" and comments and replies. People post *not* because they want to share about their lives, according to Laura. They are posting, instead, to prove how superior they are.

"It's exhausting, it really is," Laura says, referring to the competition she sees online. "I mean, it's not necessarily like, 'My Instagram is better than your Instagram,' but they want to one-up people to show, like, 'I'm in a better place than you are and I want to prove that to you and show you that I'm extremely happy,' kind of thing. Like, 'I'm doing better. I'm in a better place and I'm much happier than you are.'"

Laura makes an active effort not to get caught up in this. But she worries how much people's self-image and self-esteem are based on comparing themselves to others online. "You kind of get a false image of people and you kind of judge people based on what you see instead of really getting to know the person," she says.

And seeing all of those false "happy" images can really pull a person into a vicious cycle, Laura thinks.

Matthew, a sophomore baseball player at a mainline Protestant university, feels a lot like Laura does about people comparing themselves to others on social media—the fact of it and the frequency of it. He doesn't find this crushing personally, but he worries about the trend and its effect on his generation. People fall into this particular trap when they try to "prove" their college experience is amazing. This is something I heard over and over again from other students, too.

"Basically, you never see anyone that's not at a party," Matthew says. "It's always a party, a picture of a party or a picture of doing something cool or, you know, drinking or smoking or something like that." Matthew thinks this "warps" people's perception of what college is like—though it's not entirely misleading because these things *do* happen. "You very rarely see someone posting pictures of them[selves] studying, you know, in the library," he tells me, laughing. "It's not something that's really exciting, so it's not something they would share probably."

Everyone wants to show how exciting college is, how much fun they're having, how happy the whole experience is. It's not that they aren't happy in these moments; it's that these are the only moments they show, and that gives a distorted picture of what college is really like.

Matthew points out that the *Animal House* portrayal of college life goes way back, and that you can see this in the special collections section of his university's library. There, you can find picture after picture of the Greeks on campus from decades ago, partying and pulling pranks. But social media is highlighting what everyone has always thought—and remembered—about the college experience in a very new way, one that is far more constant and publicly accessible. You no longer need to go to the special collections section to see evidence of the pranks and the parties. Now you can click and click and click some more—every few minutes if you feel compelled. In fact, you don't even have to go looking for it; as soon as you log on, there it is, in your face, in a never-ending stream.

Even when they're not posting party pictures, people present airbrushed versions of their lives, Matthew thinks. His peers post only their successes—never their failures. They post about a vacation or a brand-new car, things that show themselves in a positive—and also financially well-off—light. Because of social media, we all seem more materialistic. People like to show off their most recent acquisitions. "Whenever somebody gets something new or nice, it's like [it] end[s] up being discussed on Facebook or Twitter." The same goes for sports, Matthew tells me. "You don't post the moments that you aren't successful in, so it seems like everyone's being successful, based on their posts. It would seem like it's easy to be successful because that's what everyone is posting. No one's posting their failures on Facebook."

What are the repercussions of seeing so much success, so constantly, on social media? It has a huge effect on us, Matthew believes—a negative one.

"I feel like maybe people get defeated a lot easier now. Like, if . . . . somebody's posted on Facebook that they went undefeated [in a sport] this season, when that other person who reads that loses a game, they're like, 'Wow, what went wrong?' Like, 'Why couldn't I go undefeated?'" Matthew worries that the stream of success on social media warps our sense of what is realistic to expect from our lives and also damages our ability to come back from failure. It saps our resilience. "If something doesn't go their way, they'll give up easier, with all the social media making it seem like success is so easy. I know a few people here that have not had great success balancing their social lives and school and so they want to transfer somewhere else. . . . Giving up on situations are almost a go-to when success isn't right there, handed to you."

In Matthew's view, comparing ourselves to those perfect, happy images has a negative effect on our overall emotional state and sense of self. "Even if you're not judging the person that you're viewing on social media," Matthew says, "you're probably comparing that person's life to your life. So if you see twelve people in a row that are having amazing times and you're not doing so hot, I'm sure that makes that person feel terrible. I know I've scrolled through and seen people doing all these fun things, and I'd be sitting at home on my couch alone and I'd be like, 'Aw, I really wish I was there. Like, why am I so lame?'" Matthew pauses at this point and backtracks a bit to think about how social media can positively affect people's happiness. "The only positive way would be from when you post something and people relate to it and 'like' it," he explains. "But I feel that probably doesn't happen near as much as when people just get on to scroll through and compare themselves to others."

Then there's Hannah, a pretty, blonde sophomore at a public university in the Southeast. Hannah describes herself as introverted. "I have also heard that I come off very intense, and that's kind of intimidating for people," she explains. "I get really frustrated in social relationships because the relationship always ends up being one-sided because I care about everything I do in my life. I want everything to be very perfect. And I think in social situations, I try to do the same thing, and I am very picky about people. So it's hard because I end up putting so much effort in and the other person ends up not putting as much effort in, and it gets really frustrating for me."

This struggle of Hannah's turns out to be a recipe for disaster on social media. She is not a fan of social media and tells me she almost never posts. If there is a major life event, she might put something on Facebook. She thinks the whole thing is really shallow. She thinks people are mean to each other, too. Hannah mentions that she was once the victim of cyberbullying—a girl from school wouldn't stop calling her a slut. Hannah doesn't want to talk about it much except to say that, looking back, she wishes she hadn't let that girl get to her, that she should have known better and had enough strength to brush it off.

Hannah wishes she didn't compare herself to other people either, but she can't help it. "This has actually been a source of conflict in my life right now because my roommate just got into one of the top sororities on campus," Hannah says. "Only the best get crowned, and they wear tiaras and call each other princesses. They have to look a certain way all the time, and so she's always done up and she's also, like, ninety pounds, not even five feet. I think she's, like, four nine. She's tiny." Hannah sighs heavily. "I'm not four eleven and I'm not ninety pounds, so I mean, I think we're all subject to beauty standards, and when she posts pictures online, she gets two hundred 'likes,' three hundred 'likes,' and I don't want to compare myself to it, but I end up doing it anyway." Being on her roommate's profiles makes Hannah feel bad about herself. She asks herself why she doesn't get that many "likes." Is it her weight, or because she's not in that ultraprestigious sorority? "And it's not just her that I do it with," Hannah admits. "I do it with all other girls."

Hannah thinks that social media is simply a vehicle for people to "show off." She even believes that it *makes* us show off, that it draws out that side of everybody. Social media is about people constantly going, "Look at me, look what I'm doing, look where I am, look at these fancy vacations I went on, look at this business trip that was paid for," Hannah says. She thinks people's reasons for posting this show-offy stuff are suspect. They're not doing it for themselves; they're doing it expressly so that other people will compare themselves and feel bad as a result. All that posting is very inauthentic, she believes.

But the thing is, all the rationalization and wishing in the world doesn't save Hannah from feeling bad about herself when she goes online. And she knows it.

ROB: LIVING FOR "LIKES" (AND WHY GIRLS GET MORE OF THEM)

"I'm a very active user," Rob tells me the moment we begin to talk about social media. Rob means Facebook, specifically—he's dedicated to Facebook above all other social media platforms. "Just walking from my class to here, I checked Facebook, and it's kind of an addiction. I don't care what other people really have to say on Facebook, so why am I looking through the newsfeed?" he wonders. But then Rob goes on to point out the benefits of Facebook. "I use it to show that I have a life, I guess. That I know people. It's kind of an ego boost."

Rob is a tall, gangly runner on his northeastern Catholic university's track team. He's incredibly nice—immediately likable—but it's also evident that he is insecure. Rob's tells me he's extremely social, to the point that he can barely get his food at the cafeteria because on his way inside he is always stuck socializing "for an hour" with all the people he knows. Rob's life centers on family and friends—they are what bring him happiness. Rob doesn't like being alone. In fact, being alone is what makes him unhappy. He mentions over and over that people are important to him—making people laugh, feeling close to others, and feeling like he's an important part of people's lives.

When he talks about using Facebook for an "ego boost," Rob is referring to status updates. "It's almost like I'd make statuses just to see how many people I can get to 'like' them, even though it's kind of superficial and it doesn't mean really anything." Rob's comment that the "likes" don't mean much seems like a reflex action, and I heard similar comments from many other students. Yet Rob is very conscious of how many he gets. The more I heard students discuss the problem of comparing oneself to others, the more it became clear that "likes" are at the heart of it.

My discussion with Rob is lengthy, and the issue of "likes" comes up often—more than in any other interview I conducted. "If I post a picture of myself at a cross-country race with my best friends, and thirty people 'like' it, if thirty people care about it, I'm like, 'Look how cool I am.' But then if nobody 'likes' it, then it's kind of like, 'Aw, I'm not that cool.' That's how I use it, to gauge how many people like you."

I mention to Rob that, while he just said that "likes" aren't important, they do sound important to him. "Yeaaaaaaah," he says, drawing out the word. "Yeah. . . . I noticed my friend, just the other day he changed his profile picture, and he called me to tell me to go 'like' it, and I'm like, 'Why do you need me [to]? You know I'm your friend. Of course I like you.' I guess he just wants to see what it looks like online. The higher the amount of 'likes,' the cooler you look; the more people think you're cool. So, I guess, *should* it mean anything? I guess I do care because I know I have a lot of friends, but it's kind of nice to see it officially written down." Rob laughs a little, nervous to admit this. He wishes "likes" didn't mean anything to him and his friends, but they do. People can't help caring about them. "It's interesting to see which people 'like' [a certain status]. If your close friends 'like' it, you're like, 'Thank you.' But then if someone you don't talk to much, 'likes' it, you're like, 'Oh, maybe I should talk to that person more,' or 'They actually care about me.' So it's kind of complicated." When I tell Rob it sounds as though he's always looking for evidence that people are paying attention, he agrees, and adds, "And I don't need that. I have so many friends to hang out with that, if somebody doesn't like me, I'm not going to let it bother me. But on Facebook, I'll post just to see how many people care. . . . It's almost, like, subconsciously just posting to see who 'likes' it."[3]

Rob doesn't really care if the "likes" are for an anonymous post either—as long as he knows it's his post, he feels satisfied. Yik Yak is enormously popular on his campus, and Rob is an active participant. People can "like" something on Yik Yak, Rob tells me, "and that doesn't give you any public awards or recognition, but you could just see, 'Oh, this many people "like" what I say or think I'm funny, and it's an ego boost. If a hundred people 'like' my [anonymous] tweet about [the food at my school], then I'll be like, 'Oh wow, people "like" or agree with me or they think I'm cool.' Like, if I were to do this in real life, they'd like me. Subconsciously you're just looking for 'likes' because 'likes' have a monetary value, it seems, even though they don't. It just seems like the more 'likes' you get, the better you feel."

I ask Rob if this affects the kinds of statuses he posts, and he hems and haws, at first saying not really, that he doesn't spend hours crafting statuses or anything. But he soon goes on to say, "If I have something to post, I'll figure out how to word it to get the most attention." I ask what he means by this, and Rob explains he wants to make sure that his post is cool or funny, as opposed to boring and straightforward. You would never simply state, "I have a cross-country race on Saturday," he tells me, because it wouldn't get many likes. But, Rob explains, if he names which schools are participating in the race and adds "Go [Northeastern Catholic]!" or "Let's get it!" the post will get more recognition. "The way you word [a post] affects how many 'likes' you get," Rob says.

Rob eventually admits that, ideally, he will only post things that get a lot of "likes." "Once I actually deleted a status because it got zero 'likes,'" he says with a chuckle. Afterward, he thought to himself that he shouldn't have posted it in the first place. If nobody "likes" something, it isn't even worth saying. "I might as well just delete it, so if people click on your profile, they could see the good stuff you posted, not something that got nothing."

This discussion of potentially deleting posts based on the response from others (or lack of one) prompts Rob to expand on his theory of "liking"—as in, what kinds of things will help a person maximize "likes"? Rob thinks he has a good handle on how the system works—and how to "game" the system. "People like jokes," he says. Though they don't just like *any* jokes, he clarifies. "If you post a funny joke that people can relate to, then it gets a lot of 'likes.' But if you post an *inside* joke, nobody will get it. Nobody will 'like' it. You have to really feed into your audience, so it's kind of like a big game . . . . like a big pointless game," he adds with another laugh.

Rob is not the first student to mention the word "audience" when referring to social media. Many students mention their awareness that they have an "audience" and that they have to be cognizant of what they post for that audience and who might think what about them. Often they are imagining future employers who might be watching for missteps. But many students, like Rob, also play to an audience of their peers.

When Rob calls getting "likes" one "big pointless game," it prompts me to press whether he really feels that way. It sounds like he made this comment because he felt that he should, even though he's obviously spent a lot of time thinking about how to "game" the system he believes is operating on social media. Rob answers with a tremendous degree of ambivalence and keeps laughing nervously as he talks. "I mean, I guess [getting 'likes' is] fun," he says. "It makes me feel better, but it really has no point. Like, what's the point of Facebook? To keep up with your friends and stay involved in people's lives that you left at home when you went to college, but other than that, what's the point of posting a status saying a joke? Who cares? I mean, when I read people's jokes, even if they're funny and I 'like' it, *I* don't care."

Rob wishes so intensely that he didn't care, but he always ends up admitting that he cares a lot. He tells me he resisted getting on Facebook until he was sixteen, which he considers almost ancient for getting an account. He was pretty anti-Facebook at first. But then he shakes his head. "And now I use it the most! I can't give you a great answer for why I do it. I don't want to say it's an ego boost, but it kind of is. It's reinforcement for acting a certain way. I like to be really social and like to tell jokes to people and like having a good time, and getting 'likes' on Facebook for that same kind of material reinforces acting that way in public around those people."

Our interview soon moves on to other topics, but it isn't long before Rob cycles back to his theory on the centrality of getting "likes" and also which posts—and who—get the most "likes" of all. Whenever Rob thinks of making a post, he wonders to himself, is it worth saying? In other words, will people "like" it? If his answer to this question is no, then, "It's not even worth putting on there," Rob claims. "There's no point in saying, 'Oh, the sky is blue.' No one's going to care, so why write that down? I try to do stuff that will get positive feedback or at least get some 'likes.' "

Rob feels the deck is stacked against him in the competition for "likes." Women always get more "likes," he thinks. His frustration about this is evident. He thinks it's mystifying and also unfair. "A girl just posts a random picture in her room laying down on a bed, and it gets a hundred 'likes'! . . . .I think it's interesting that some girls don't even need to try just because they're female." Rob mimics them, saying, "Like, 'I can lay on my bed and take a picture!'" and rolling his eyes. He envies girls' ease in deciding what to post.

In theory, Rob can accept that he's not going to get "a hundred 'likes,'" but he cannot accept the disparity of "likes" by gender. "I actually go on a *cruise* to Bermuda, you know, the clear water ocean, on a cliff, with the water in the background—it was a *cool* picture. It gets thirty 'likes.' Girls standing in a dark room, sitting in a dark room on a bed, just reading a book—a hundred 'likes'!" Rob thinks the disparity has to do with "sexual stuff." Guys like pretty girls, and when guys go on Facebook, they'll simply "like" girls' pictures but won't "like" the pictures of other guys as much. So guys start off with a disadvantage in the "likes" department. But Rob also suspects that girls are gaming the system, too, and are better at it than guys are.

"I think a lot of girls re-upload their pictures," he explains. "Like, I post it once, and then after a certain time, people stop seeing it because, unless they go on my profile, it doesn't pop up on the newsfeed. While girls sometimes upload their picture and then, later on, they re-upload it, so it shows up again but the 'likes' [they got from the first upload] stay there." How does Rob know that "likes" stay attached to the photo even if you upload it again? Well, he tried this once himself, he admits *very* sheepishly. Actually, more than once. He got a couple more "likes" but did not see the kind of success that he thinks women achieve when they employ this strategy. Rob thinks women do this all the time. He's also bitter about his suspicion that if he changed his relationship status to "In a relationship" on Facebook, he'd only get "five people" to like it, whereas a woman would probably get "a hundred 'likes.'"

Rob's worst fear seems to be getting no "likes" at all. The thought that one of his posts would be completely unimportant to people fills him with shame, which is why he's spent so much time theorizing about which posts get the most "likes" and why.

Overall, Rob feels like social media is "a mix of a popularity contest and an ego boost contest," he says. "Like, who can get the biggest ego on Facebook." Rob thinks, for instance, that people will accept any friend request, just so they can increase their number of friends. He thinks people don't care if their friends on Facebook are "fake friends" because all that matters is "looking more popular." Rob is sounding more and more cynical as we speak. But then he lights up and tells me he has eight hundred friends on Facebook, and that he accepts random friend requests from people because he "might as well," he says. "It will look better if I do than if I don't. But in reality it doesn't really matter. Well, I guess it matters to me," he adds, with that now familiar nervous laugh.

When Rob gets a lot of "likes" on Facebook, he says, "it just confirms your beliefs about yourself." If people affirm his posts with "likes," it helps him to believe that the things he does in life (cross-country, for example) matter. When he doesn't get "likes," it confirms which things *don't* matter. In other words, Rob judges his actions, his choices, his endeavors, and even his life goals based on how others affirm—or do not affirm—those same things on social media. He derives a tremendous amount of meaning from what others tell him about his life through "likes" on his posts. His stress about how much "likes" mean to him is painfully evident from the number of times he mentions it. He doesn't understand why they're so important to him, though, and he wishes with all his might that they weren't. But they *are*, he always confirms. He mentions on six different occasions during our interview that "likes" give him a much-needed "ego boost."

What is it about Rob that makes him care so much about this kind of approval, whereas other students—many of whom are not nearly as successful socially as Rob claims to be—truly could not care less? It seems that a steady stream of social media affirmation helps to remind Rob that he exists (in general), but also that he matters socially and that his ideas and updates are worth paying attention to. People "like" what Rob says; therefore he *is*.

HIGH TRAFFIC TIMES, AND MY BEST-SELLING AUTOBIOGRAPHY

Should we worry about young adults, like Rob, who obsess over this aspect of Facebook and other social media platforms? Or is this all harmless, a passing fad that won't mean much as Rob gets older? Rob isn't alone in how he feels; many of the students I surveyed had similarly intense views and worries about "likes," concerns that defined much of how they operated and what they posted online. They, too, stressed about "likes" even as they also *really* longed not to notice whether or not they got them.[4]

I asked everyone I interviewed what they thought of "likes" and whether they cared about getting them. Some students were adamant that they didn't care at all, that they barely thought about "likes." Some went so far as to say that they really hated the whole "likes" business, just as they do "retweets" and "favorites" and anything else that people can count up. Many students mentioned how awkward and weird it is when people "like" posts about someone being sick or sad, or having a family member pass away. (Why would you "like" that? they wondered. Isn't it inappropriate to "like" someone's grief?)[5]

But just as many students grew sheepish at this question, rolled their eyes, and said something about how, while they wished they didn't care about "likes," because they are superficial and therefore shouldn't matter, of course they cared! How could they not?

For example, Mercedes marvels that there even was a time in her life when "likes" didn't yet exist. In 2006, the year she first got onto Facebook, "likes" were not yet a feature, and she almost can't remember life without them affecting every comment, update, and photo a person posts on Facebook. "I think whenever people post things, they are hoping for some kind of reaction," she says. "I mean, if I posted that I'm 'In a Relationship' on Facebook, I'm going to be hoping that people would 'like' it." In fact, a month before our interview, Mercedes and her boyfriend did just this, and they broke the "100 likes" barrier (most students seem to consider the first time a person gets a hundred "likes" as a kind of Facebook milestone). She was pretty happy about this, though even as she tells me that she cares she also chuckles and rolls her eyes. "You're like, 'I feel so good about this!' But you're [also] like, 'Why?' You should feel good about the relationship! Why did having more 'likes' make it cooler?" Mercedes goes on to talk about how getting all those "likes" definitely gave her a rush, but she felt "stupid" that she succumbed to it. She is quick to say that "likes" "mean nothing" and that it's "dumb" that they make people feel anything at all.

People judge their profile pictures based on how many "likes" they get, too, according to Mercedes. Unless you get at least 30 "likes," the picture "doesn't count, it doesn't count. You should change it right away. That's how people think." I ask Mercedes if this is her rule of thumb about profile pictures. She says no, she wouldn't take a picture down just because it failed to get the requisite thirty "likes." But it does make her doubt the picture. She may have thought a picture was cute, but the low number of "likes" tells her she was wrong.

This is when I first hear about "high-traffic times" for posting. The idea is to post at certain times of day to maximize the number of "likes."

"A lot of people would say, and my friends would tell you this too, it depends on the time of day that you post this information—that's how crazy it is," Mercedes explains. "If you change your profile picture at seven at night, you're going to get more 'likes' because more people are on Facebook at seven at night. . . . But if you change your profile picture at two in the afternoon, then you're not going get as many 'likes.'" I ask Mercedes if she follows these guidelines. Not necessarily, she tells me. But she certainly has many friends who do.

Like so many students, each time Mercedes posts and doesn't get a lot of "likes," she goes through a kind of rationalization process to console herself. When she feels the disappointment coming on, Mercedes reminds herself that "likes" don't really matter. "This is *my* page," she tells herself. "So it's just about things that *I* like. [It doesn't] have to mean that everyone else will like it."

Yet, for students like Matthew (mentioned above), gaining approval and getting "likes" can become a constructive and positive part of one's self-understanding. Matthew tells me about "My Story," a feature on Snapchat that compiles photos from a single day, and the Timeline feature on Facebook. Matthew's reflections intrigue me. "I feel like everybody wants to have a story," he says. "Everybody wants an autobiography at the end of their life, a bestseller, and [My Story] is a way to do it now . . . .you know, a way to get that feeling for a moment or two. Like, 'Hey, check this out! This is me. This is my life! Look at me!'" Matthew pauses a moment, then explains that before our interview he'd never thought of his collection of posts as an "autobiography," but now that he's said this out loud, he thinks it makes sense. "I'm pretty sure there's a picture or post about almost every big event in my life on Facebook or on Twitter somewhere," he says. "Snapchat is more of a day-to-day experience [because snaps disappear], but Facebook and Twitter definitely, on some level, become your autobiography." The Facebook Timeline itself, Matthew explains, is a kind of autobiography because it captures the significant things that have happened throughout your life. That is its purpose.

Matthew goes on to equate posting something that gets a lot of favorites, "likes," or "retweets" as "sort of like saying your autobiography is a bestseller"—at least for a moment. Having so many people see whatever it is you've put up about yourself, and then approve of it, letting you know that they've seen it and "liked" it, shows you that everyone can relate to whatever you've just posted. He gave me an example. He once sent a photo via Snapchat that tons of people took screenshots of because they loved it so much and it made them laugh. He told me how much he loved that so many people saw it and enjoyed it and how this made him feel appreciated. Matthew thinks that people are kidding themselves if they don't want their posts to be favorited, retweeted, and "liked"—he certainly wants it, and says that getting "likes" is one of the main incentives to post.

Then there's Maria, who at one point abandoned her Instagram page because her photos weren't getting any "likes" from friends, which made her feel terrible. "It's so easy to press 'like,' and nobody was 'liking' anything," she says. Maria felt especially bad when photos that were important to her went unnoticed by the people she cares about. But then Maria created a new Instagram account exclusively for pictures of her dog—"people love pictures of dogs," she says—and now she gets tons of "likes." Maria swears she didn't do it just to get "likes," but she admits she loves having so many unexpected fans, and that this makes updating her Instagram page "more fulfilling."

Maria says she no longer actively searches her Facebook page to see if people are "liking" her photos. She doesn't love the way Facebook operates—what things stay on top of the feed and why. "So, [my friend] got 170 'likes' for this one picture, and it was there for days at the top [of the feed]. So then I 'liked' it, because there's so much exposure, but that's an interesting thing: Facebook decides to expose you to something based on how they feel it ranks [and how] it should rank to you." Maria thinks this is "weird and creepy," but she also doesn't want to miss the things that are popular on Facebook, so it can be a useful tool. Maria doesn't judge the "success" of her posts by this, however. For example, she tells me, "There's a picture of me and my old dog [on Facebook] before he passed away, and we're hugging and it's kind of an artistic photo and only a few of my friends 'liked' it." She adds with a laugh, "But I didn't care. I thought it was great anyway. And it was a great success to me."

You always get a response from people if you post during "prime time," according to Maria. "You know, like six, seven [o'clock], when people are getting done with class and they're not doing anything," she explains. This is when you can expect people to be online and when you can expect the best—or, at least, the widest—reaction from people on your friends list. Of course, if you post something then and you don't get any reaction, it can make you feel even worse.

There is a tried-and-true way to ensure you get a lot of "likes," though—well, a tried-and-true method for women. While Rob complained that girls can take just about any shot of themselves, put it up online, and get a ton of "likes," Maria sees something slightly different—and incredibly worrisome—going on. Generally, the photos that get the most "likes" are provocative photos of women. Maria tells the story of someone she knew from home who had a photo on her page that got more than two thousand "likes." The photo was very revealing, she explains. "It was disturbing, and it was also scary because she kept posting more and more of those photos, and they were getting a ton of 'likes,' " Maria says. "And it was kind of upsetting to me to realize that she kept doing this for the popularity of it. The photos were of no one except for her in a bikini or doing something suggestive and then that's all her profile is now. It's thousands of 'likes' and those kind of pictures. It made me sad."

Be careful, though, Maria warns. If you post too many of those suggestive photos, eventually people will get bored and start ignoring them, and you won't get that much-desired boost. Even people on Facebook have limits. This happened with Maria's best friend. "She did that kind of thing and posted all of these pictures and they were getting 'likes' for a long time, like these provocative photos," Maria says. "I think people started getting sick of it or something and then, out of nowhere, people just stopped 'liking' them, and eventually she stopped posting altogether because nobody 'liked' the stuff." It's as if people had built up a tolerance.

"Likes," "retweets," "shares," upvotes and downvotes—all of these are methods for viewers to actively judge (or render invisible, irrelevant, and uninteresting) users' content. These functions can provide affirmation (when you get enough of them, and they are positive), but they also raise the stakes of posting and heighten everyone's sense that they have an audience out there watching and evaluating them. Most students try not to care about the reaction they get when the post, but the fact that there are mechanisms people can use to show you that they care—or don't—is nearly impossible to ignore. Posting on social media is rarely ever innocent. You don't post simply because you feel like it or because you're just employing your right to self-expression. You post with at least the slight hope—if not the profound one—that everyone will see your post and respond positively. No response or a negative one can render you miserable. "Likes" take the usual highs and lows of young adult social life and make them quantifiable—a running tally of your self-worth.

Try as students may, most of them can't seem to *not* care about these things.

Then I meet Avery, a junior at a southern university, who thinks that social media can be a competitive place—in a "subconscious way," she claims. "People are just trying to put flattering things about themselves online, and then, everyone else is trying to keep up. That makes people competitive."

Avery and I have a long conversation about the kinds of things you should post—and shouldn't post—to maintain a positive appearance. But something Avery says strikes me as particularly revealing of the plight I keep hearing described about being subjected to how *awesome* and *amazing* everybody else is doing, seemingly *all the time*—this constant stream of "Look at me! I'm so happy!" that can affect people emotionally in negative ways. As students compulsively check their accounts all day, every day—when they wake up, at breakfast, when they're walking to classes, when they're in class, while they're with their friends, even while they're at sports practices and games, certainly before bed and sometimes in the middle of the night—it pains and exhausts them.

"I think it can damage self-esteem because you're comparing everything about your life, the good stuff and the bad stuff, to *just* the good stuff of everyone else's lives that you see online," Avery says, reminding me of Emma's similar remarks.

A strange kind of relational triangle can develop as a result. When someone goes to check in on everyone else, they bring all sides of themselves to the table as they are scrolling and lurking—the happy, secure sides, but also the unhappy, the struggling, and sometimes the depressed sides. But you see only the happy side of everyone else, "and then you feel really crappy in comparison," Avery says. Being confronted with so much joy and showing off can send your self-esteem into a spiral. Avery thinks that when people go on social media, they often forget that the people whose posts and photos they are looking at also have "bad stuff" going on, and that they're just not showing it. "I think [the people scrolling need to] talk themselves into realizing that everyone has good stuff and bad stuff going on in their lives, but people don't normally stop to filter through everything they're thinking, they just think it and feel it. They don't reason."

And that can really get a person down.

COMPARING OURSELVES TO OTHERS: A KIND OF SICKNESS WE SUFFER

We don't need social media to compare ourselves to other people or to worry about others' responses to what we say or do. Engaging in status competitions is nothing new. It's common, especially when we're young (though adults are certainly not exempt). We compare our looks, our hairstyles, our opportunities, our friends, our successes and failures, where we've traveled (or haven't), where we've gone to school, where we're from, our clothes, and all sorts of material objects. The list goes on and on. We seek approval and affirmation all the time.

The difference with social media is that it seems expressly designed for this purpose of showing off, for bragging and boasting of all that one is, has, and does, as well as for others to judge these things. Facebook is the CNN of envy, a kind of 24/7 news cycle of who's cool, who's not, who's up, and who's down. In the process of showing off our successes and proving how happy we are, we are exposing ourselves to others' highlight reels, too, and to the possibility of rejection. Unless you have rock-solid self-esteem, are impervious to jealousy, or have an extraordinarily rational capacity to remind yourself exactly what everyone is doing when they post their glories on social media, it's difficult not to care.

The title of the actress and comedian Mindy Kaling's first memoir, *Is Everybody Hanging Out Without Me? (And Other Concerns)*, not only reflects her trademark humor, but it aptly captures this common pain, so much so that I borrowed it for this chapter.[6] There is even a special acronym to capture the phenomenon of so much comparing ourselves to others today: FOMO (Fear of Missing Out), a kind of suffering that comes from witnessing the amazing times other people are having without you, and perhaps have intentionally *not* invited you to join.[7] Some of us are really good at protecting ourselves from such comparisons, or at least being ambivalent about them, but many of us are vulnerable to this particular suffering, and social media exacerbates and escalates it to levels that most of us have never before experienced, nor have the emotional resources to withstand—at least not so consistently and constantly.[8]

There are benefits to this, though. All those "likes" and "retweets" and "shares" can make us feel like mini-celebrities, at least for a moment.

Before social media, we may have sat in a classroom admiring someone's stylish outfit, or wishing we had a cute boyfriend, too, or that we had gotten to go to that fun party on Friday night. But there was only so much we could see and hear about. With social media, people can log on again and again, stare and envy and obsess as long as they like, and not only *know* that someone has a cute boyfriend but peruse all the gorgeous pictures of him, not only *hear* there was a party on Friday but see the hundreds of photos of people having a blast without them.

In the online survey, students were asked to respond either "yes," "no," or "not applicable" to a series of statements about how they feel/are/act/what they do when they go onto social media. One of those statements was "I find myself comparing myself to others a lot." Of the students who responded, 55 percent said yes and 43 percent said no (2 percent answered not applicable).[9] Well over half do this type of comparing regularly, but when it came to responding to the statement "It's important that others see me as having a good time," only 35 percent of these same respondents said yes, and a whopping 60 percent answered no.[10] So while the majority indicate that they often compare themselves to others, at the same time they do not believe they are "showing off," as Hannah might put it. Even though many students complained during the interviews about all the boasting, bragging, and competition they see on social media, and the one-upping that Laura spoke so much about, only about a third of students in the online survey admitted that they actively care about others perceiving them in a particularly good way. This disconnect is apparent when when Rob talks about "likes." He is obviously obsessed with them, but he doesn't want to admit that he is.

In an optional open-ended essay question that asked students if they ever find themselves comparing themselves to others and/or feeling that they're missing out on something, just over half of the 320 students who chose to answer said they feel this way. The tone tended toward the exasperated, with answers such as, "Well, *obviously* I feel these things." One reply even began with "Duh!" Students who answered yes often added a comment such as "This makes me sad" or "It usually makes me feel bad about myself." Not only does it make them upset, but the fact that they care makes them feel even worse. "I wish I didn't do this," one wrote. "I try not to do this but I can't seem to help myself," added another. Students assess the behavior as depressing, upsetting, or a "bad habit"; they indicated that it makes them feel "pressured," inferior, behind, or "not pretty enough," and that it "messes" with people.

Only fifty students (16 percent) claimed that they'd never compared themselves to others or felt left out. Even those who said "I used to do it, but not anymore" talked about how they "got themselves over" this tendency, as though they'd healed themselves of a terrible habit, a kind of sickness that comes with social media. Some students suffer far worse than others. When students talk about feeling inferior on social media— even if only rarely or in their past—they discuss it as if it is an affliction, one that comes with the territory. And a number of students commented on how comparing oneself to others and having feelings of being left out are simply human tendencies that have been heightened to an unhealthy degree by social media.

As much as so many students try not to succumb, the phenomenon of comparing oneself to others—and the feelings of missing out that come with this behavior—seems to be one of the most common experiences of being on social media. Getting "likes" (and "retweets" and "shares") is a central part of the performance of perfection and positivity. Not only does it prove that you "are," as Rob might put it, but it also is a quantifiable mark of success and affirmation (part and parcel of that "popularity principle" that Van Dijck talks about)—it confirms that you are performing your social media duties well. On the flip side, the inability to obtain quantifiable public approval is a source of shame. It shows everyone not only that you are irrelevant, but also that you are performing poorly in the endeavor to showcase only your best efforts, your greatest successes, your most attractive traits, both physical and personal—that even when you are making your best effort, you still can't measure up to the competition. For especially sensitive and vulnerable students like Margaret and Michael, the only option seems to be fleeing social media altogether, or at least nearly, so they can hang on to some semblance of emotional well-being.

Of course, the pressure students feel to post positive things about themselves only serves to heighten the experience that Emma so vividly described as comparing the worse version of herself to the best version of everybody else. What's more, students not only are acutely aware that they are being judged every minute of the day by their peers but they also know that they might be judged later on by people with even greater power over them: potential employers.

2

THE PROFESSIONALIZATION OF FACEBOOK

(AND WHY EVERYONE SHOULD KEEP THEIR OPINIONS TO THEMSELVES)

I have heard a lot about being professional and trying to avoid any pictures that may jeopardize obtaining a job. Since I am a college student in a global economy along with a recession, I am well aware of the little details and how they can be a deal-breaker for my postcollege life.

Branson, junior, Catholic university

I've been briefly coached on being careful about posting negative things and how such negative posts will reflect on my personality to potential employers.

Ari, first-year, private-secular university

As an athlete, we are given strict instructions on posting. They teach us about our personal brand and image, and how one negative post can have lasting impacts [sic].

Isabel, senior, evangelical Christian college

AAMIR: MAINTAINING A SPOTLESS RECORD

Aamir is a tall, dark-eyed first-year student at a private university in the Southwest. He's sincere and direct, and he approaches our interview as he might an important exam. Aamir is a political science major, and it isn't long before he tells me that a political career is in his future. In fact, it's nearly the first thing he mentions, and he mentions it repeatedly. He talks about the importance of his community and his "constituents" as though he is already a seasoned politician. Aamir is very active both on campus and online—he's on Facebook, Twitter, and Instagram—and he keeps his political future in mind every time he posts (or decides not to). In fact, Aamir's teachers have been preparing him for years to put his best foot forward by making sure that his behavior on social media is "spotless." Future politicians can't afford any missteps—not even while they're in middle school.

Aamir is one of the few students I meet who posts about politics (he's a left-leaning Democrat, he tells me), but he's still very careful about what he says and where he says it. He will post about politics only on Twitter. "Most of my tweets are pretty left-leaning, and I don't think that's going to change for a very long time, so I'm not worried about posting things that are honest ideologically," Aamir says. "But I just don't write a post that might make me seem irresponsible. So, for the most part my tweets are pretty dry." For Aamir, the revelation that Mitt Romney bullied a classmate in high school is enough of a cautionary tale to warn him off unseemly behavior—or, at least, leaving a record of it. "If you post stuff like that on social media, it's permanent," he comments. "The risk that, you know, someone could look at it in the future and then take that edge as your character is far greater. So I think that the world we live in now is a lot more dangerous especially with social media." For Aamir, "danger" and "risk" on social media have nothing to do with bullying or predatory behavior. They are about public perceptions of one's character. That Aamir goes so far as to perceive social media as a "dangerous" force because of the threat it poses to a person's reputation is indicative of his intense fear of making any missteps. One post could cost him his dreams. "I'm sure there will be a lot of folks that run [for political office] in the future," he says. "Or probably that are running now, that are right now posting things that are, you know, less than favorable to their image. But, you know, I try to have as spotless of an image as possible."

I ask Aamir if he has always tried to have a "spotless image" on social media, and he informs me that his efforts started back in middle school, and that "spotlessness" was *mandated* in high school. "I became a part of student council in my high school, and our high school's student council was extremely strict, especially if you were in leadership, with the stuff that you can put on social media," he explains. "They had a very similar philosophy that the leaders should have, you know, spotless images on social media at home or wherever you are. So I think that really taught me how to avoid certain things and just not put it on social media."

Aamir learned this from one of his advisers—and from the mistakes his peers in student government made. If a member of the student council posted something improper, in addition to forcing the student to take down the offending post, the adviser took disciplinary action. The policy "was pretty strict, and it was definitely pushed," according to Aamir. It wasn't only the advisers that pushed it, either. "Sometimes the other students would try to keep you on track or report you if you didn't do what you were supposed to," Aamir says.

There are peer enforcers everywhere now—in sororities and fraternities, certainly, but even in student government. The Greeks are trying to protect the chapter's reputation on campus so they can continue to party however they want, whereas Aamir's fellow student politicians are trying to catch each other out because they're in competition. Just as we see presidential candidates playing up an opponent's errors and missteps for all they're worth, these students stand to benefit from the mistakes of their peers. The public nature of social media offers lessons in these tactics to the very young and politically minded.

Aamir thinks this was excellent preparation for his future. He's grateful to have gotten such social media training at an early age—he feels like he's dodged a bullet, and that it gives him an edge over peers who haven't been as careful for as long. Aamir is always thinking about the competition he faces, and social media offers him a constant reminder that competitors are always ready and waiting to steal the spotlight if they can. He wasn't always so careful online, though.

There *was* a time, early on in middle school, when Aamir used to swear and make vulgar jokes on social media. I ask him if he's done a

"Facebook Cleanup" of these unbecoming remarks—something that many college students tell me they've done—and I'm surprised when he tells me no, he thinks they're still there. Because Aamir was so young (around seventh grade), he thinks if people found those posts now, they'd give him a pass. "I feel like it's excusable," he says. But then he looks at me intently and asks, "I mean, now that you've mentioned it, is there a way to clean stuff up?" I explain that there isn't a program you can download as far as I know—I've just heard other students discussing their Facebook Cleanups.

Aamir seems slightly disappointed.

After his teachers started helping him pay attention to how he operates on social media, Aamir developed a set of rules for his posts: no swearing, nothing overtly sexual, and no posting about emotions—unless they're positive ones. Aamir limits his political posts to Twitter because he feels that Twitter is the place where political posts are the most accepted, whereas on Facebook you have to be far more careful about posting anything provocative. Also, Aamir is running for office in his university's student government this coming spring, and most students from his school are on Facebook. "I try to avoid posting things on Facebook because it might, you know, blow up my chance of getting elected," he says. Aamir's college peers have yet to find him on Twitter, so he feels safer being himself there. He is constantly engaged in a dance of minimizing "risk" and avoiding "danger."

Despite the possible pitfalls, and how difficult it can be to maintain that spotless record for years on end, Aamir knows it's important for a future politician to maintain a presence on social media. He believes President Obama succeeded because of it. "Having a good social media presence and being able to use it well getting your message out," Aamir says, "I think is definitely a very useful skill to have especially nowadays."

As long as he keeps things clean.

THE CAREER-DESTROYING FACEBOOK POST AND THE POLICING OF YOUNG ADULTS' SOCIAL MEDIA ACCOUNTS

Not everyone is as vigilant as Aamir, who takes caution to the extreme. But this careful attitude about what to and what not to post was nearly universal among the students I interviewed. They believe that all sorts of people are scrutinizing their social media accounts—that everyone is constantly being watched and judged and potentially shamed for mistakes.[1] Their biggest fear is that potential employers will find an offending comment or photo on one of their social media accounts. They aren't crazy to think so, either—employers really do check the social media profiles of applicants.[2] Because college admissions officers do the same, many students have already gotten a taste of this variety of online stress before even arriving on campus—and once they do arrive, the scrutiny only continues.[3]

On the most sinister level, this sounds a lot like the interpretation of Bentham's panopticon in Foucault's *Discipline and Punish*, but on a virtual scale. College students (and young adults in general) are highly aware that because of social media, they can be "spied upon" at any and all moments by people they've never met, can't see, and who may hail from far-flung locations, all of whom may have power over their lives. To draw on Foucault, this makes participants on social media "inmates" of a virtual sort, because no one knows "whether he is being looked at at any one moment; but he must be sure that he may always be so."[4] Young adults know (by way of the adults in their lives who've warned them) that this is effectively the deal they've signed up for on social media—by attaching their names to various platforms and by commenting and posting in these public spaces, they are allowing themselves to be surveilled and policed on a constant basis and must behave accordingly or suffer the punishment of not getting a job after college, or not even getting *in* to college. In this same vein, Daniel Trottier has written extensively on the relationship between social media and surveillance, and devotes consideral energy to investigating how university administrators have come to regard surveilling their students' social media activity (especially on Facebook) as a new and important addition to their professional responsibilities—important both for the personal safety and future prospects of their students, but also to the well-being and reputation of the institution they represent.[5]

But the notion that we are using social media to police, spy upon, and subsequently judge our children, students, and future employees (however well-intentioned we might be in doing so) is rather chilling. In *It's Complicated: The Social Lives of Networked Teens*, danah boyd alludes to a generational conflict around the idea of privacy, and that even posts that teens make in public forums like Facebook deserve to remain private from certain constituencies (namely parents), much like the diaries of old that teens kept to record their most secret thoughts and feelings. The widely accepted notion that all young adults today overshare and therefore don't care about privacy is wrong according to boyd, and she argues passionately that we've misunderstood teens' relationship to privacy—they do desire it, intensely so, and believe that it is inappropriate for "parents, teachers, and other immediate authority figures" to surveill them on social media.[6]

But the college students I interviewed seem to have given up on the idea of retaining any semblance of privacy on social media forums attached to their names. They've accepted the notion that *nothing* they post is sacred and *everything* is fair game for evaluation by anyone with even the littlest bit of power over them—that these are simply the cards their generation has been dealt.

Hence, the Facebook Cleanup that so many of the students talked about at length, has sprung up in response to the situation that young adults find themselves in today. A Facebook Cleanup involves going back through one's timeline and deleting any posts that do the following:

- Show negative emotion (from back when you didn't know any better)
- Are mean or picked fights
- Express opinions about politics and/or religion
- Include potentially inappropriate photos (bikini pictures, ugly or embarrassing photos, pictures showing drinking or drugs)
- Make you seem silly or even boring
- Reveal that you are irrelevant and unpopular because they have zero "likes"

Students who aren't quite sure what to keep or delete can consult numerous news articles for advice.[7] The term "Facebook Cleanup" is apt, since students' fears are very much focused on Facebook rather than other social media sites. Facebook is so ubiquitous among people of all ages, that students consider it the go-to app for employers. As a result, it's the social network where they feel they need to be most careful, and where they believe the creation of what they refer to as the "highlight reel" is most essential. For college students, the highlight reel is a social media résumé, usually posted to Facebook, which chronicles a person's best moments, and *only* the best moments: the graduation pictures, sports victories, college acceptances, the news about the coveted, prestigious internship you just got. Students often organize their Facebook timelines to showcase only these achievements and use other platforms for their real socializing.

In *The App Generation*, Howard Gardner and Katie Davis briefly remark on the notion of a "polished," "shined-up," "glammed-up" profile, "the packaged self that will meet the approval of college admissions officers and prospective employers."[8] Students are indeed seeking control of public perception in a sphere where loss of control is so common and so punishing. Gardner and Davis worry that young adults' belief that they can "package" the self leads to a subsequent belief that they can "map out" everything, to the point where they suffer from the "delusion" that "if they make careful practical plans, they will face no future challenges or obstacles to success."[9] This makes everyone disproportionately outcome-focused and overly pragmatic, especially in terms of careers, and less likely to be dreamers.

The students I interviewed and surveyed certainly *are* outcome-focused in their online behavior when it comes to pleasing future employers. But rather than giving the students a false sense of control, which Gardner and Davis find so worrisome, this focus seems instead to make them feel incredibly insecure.

And who can blame them?

They are acutely aware of the precariousness of their reputations and the potential that their lives could be ruined in an instant. The students I spoke to clearly understand that they *lack* control because of the way they're constantly being monitored online. This knowledge was central to nearly every conversation I had.

Students consider a couple of topics particularly treacherous. One is politics, a subject on which they believe they can easily and quickly lose control if they aren't careful.[10]

One young man's social media filter is so well-developed that he tells me he would never post anything provocative, for fear of the possible repercussions. "I'm not one of those people that just goes and posts their social opinion on the gun policy or gays being able to get married and then starting a whole bunch of arguments on social media where anyone can see it," he says. Photos that include alcohol are also off limits, even if you are over twenty-one. He's made sure that anyone who goes to his Facebook page sees that he is outgoing, popular, and active in various campus groups, and that he "seems like a respectable person." He's learned to use his Facebook account more like an online résumé—the highlight reel I hear about constantly from students—for the benefit of future employers. He finds it hard to believe that some people haven't learned this lesson yet. "I hate when people post their political or social ideas," he tells me. "Once it's out there it's always out there. You don't want a future employer being able to see that and then them just getting an instant, maybe negative image of you, because they have a different view. So, I just try to stay away from that. I just try and stay neutral in everything on social media."

Even George, a conservative gun enthusiast who attends an evangelical Christian college, worries about the fine line he is walking between connecting with like-minded peers and potentially offending future employers. George primarily sees social media as a vehicle to connect with other gun enthusiasts and advocates of Second Amendment rights. "I've always been into firearms—I have since I was a kid," he tells me. "I'll post a lot of Second Amendment stuff. I'll post a few Republican posts here and there for certain politicians if I know them or if I'm a big advocate." He feels somewhat "safe" in posting about the Second Amendment because all of the people he's connected to online are also gun enthusiasts.

However, when we discuss the problem of future employers reviewing people's social media profiles, George echoes his fellow students. "You don't want to put on there you being held upside down by your ankles for a keg stand," he says. "And don't put stupid stuff on there." George knows that his political views can get him into trouble if he's not careful—the gun-related posts especially. "I don't put anything that's super left- or right-wing," he tells me. "Stick to fairly moderate, big issues, because you don't want to offend people. If you get an employer that is a Republican or is a Democrat, and their views conflict with yours, even though they shouldn't care, they're going to be biased. Everybody's biased and everybody's selfish. That's a basic economic principle. Everybody is self-interested. If they got two guys there, you're exactly the same, and your employer's a Democrat and you've got one [candidate] that's a Republican and one that's a Democrat, then the Democrat stands a better chance of getting [the job] because [the employer] is self-interested. It's not necessarily that [the employer] is trying to be biased, it's in his state to go with somebody that's more like-minded."

Appearing happy and positive in order to please potential employers was also of central concern to students. In the online survey, students were asked to reply yes or no to the following statement:

I worry about how negative posts might look to future employers.

Seventy-eight percent of the respondents answered yes. And, when asked about the dos and don'ts of social media, 30 percent said that the main advice they've received about social media was related to employment.[11]

Most students feel that they are in a "damned if you do, damned if you don't" situation. They worry about whether future employers might find an offending photo or post attached to their names. But many of them also worry about what happens if they have no social media presence at all for their future employers to scrutinize and dissect. If they are completely absent from social media, this might seem suspicious, too—a future employer might assume the worst about them (i.e., their photos and posts are so vulgar and unbecoming that they've hidden everything from the public). As a result, many students keep their Facebook pages updated with the highlight reel—a G-rated social media résumé that employers are welcome to peruse. As long as employers can find *something*, and that something is good, students believe they've done their social media homework.

THE GOOD OL' DAYS

There was, among the students I met, one exception to the rules about Facebook Cleanups and what future employers will likely forgive. The kinds of roll-your-eyes embarrassing, unbecoming, ridiculous things a person puts up during middle school—when the thrill of getting your first social media account is overwhelming—are often given a pass. Even Aamir, fervent advocate of the "spotless record," feels this way (mostly).

Many students believe these sorts of comments and goofy pictures (even the occasional mean post) can stay because they were just kids when they posted them, too young to know any better. What's more, some students are nostalgic about those posts and like to go back and laugh at their younger, sillier, more uninhibited selves.

Avery, a senior, regards those early posts with affection. "I like looking back to what I used to post, old photos I'm tagged in and stuff," she says. "I think that's kind of fun. . . . I think that a lot of kids my age will go look [back]. Sometimes a really old post that they made will resurface. . . . like a picture that we think is funny now, but they thought was serious at the time, and some really deep quote or something like that, which they're embarrassed by now."

But once you hit high school, all of that has to change. As you get older, you learn how *not* to post like that anymore.

"Starting in ninth grade, I did a big Facebook Cleanup before I went to college," Avery says. "I just figured I needed to get rid of the embarrassing pictures, the embarrassing selfies from ninth grade and stuff like that." Earlier than ninth grade, though, can stay.

In fact, many students view the shift in their habits in a way that calls to mind the Garden of Eden. There was a time of innocence, when they would post whatever they wanted, and then they learn that people will use their online activity against them, and they begin to feel shame. They see the transition from one state to the other as a new kind of rite of passage from childhood to adulthood. There is a fondness in their voices when they harken back to those early days, when they weren't so aware of the need to appear perfect, pandering to future employers (or college admissions people or coaches or sorority sisters). They are remembering a time when they could simply *be themselves* without penalty or punishment—at least not a lasting one.

One young man tells me he did a "complete one-eighty" with regard to what he posts and doesn't. He talks specifically about how when he was younger his posts were more immediate and less filtered. "It's more of a mature level that I operate now than compared to when I was young," he says. "If I had a thought in my head, *boom*, to Facebook it went. To Twitter it went. If I saw something in the news that I thought looked interesting, or I thought would generate a discussion, *boom*, to Facebook it went."

Now that he is older, he stays away from anything that might be deemed even a little bit provocative.

Certainly by college, those carefree days of social media are gone for good. Those who have grown up on social media are professional about it. And to be professional about it means, for most students, maintaining the facade of a happy, successful life that will impress future employers.

THE PEOPLE BACK HOME

The message about professionalism on social media seems (at least for now) to be focused mainly on college students. It is relentless and unchanging. One of the only meaningful distinctions I noticed was *when* students began to feel this pressure, *when* they learned to project their best selves online. College students like Aamir, who hail from wealthy families or communities where everyone is college-bound and groomed for it from a young age, may learn this lesson as early as middle school, whereas students from poorer backgrounds where going to college is not a given for everyone (or anyone) may only learn this *after* arriving at college.

But once they are on campus, the message is drilled into them.

The more prestigious the school, the earlier the students seem to have learned about Facebook Cleanups. Most colleges now offer training on how to appear upstanding online—refreshers for those who know this already, and wake-up calls for any stragglers. But students at the most academically rigorous and wealthy institutions tend to absorb this lesson in high school.

It is on my visit to a public university in the rural Midwest that I first notice the relationship between economic background and the point at which people learn to scrub their profiles. The students at this institution come largely from either tough inner-city neighborhoods, from which few people make it to college, or from some of the most rural areas in the state, where their parents are struggling farmers or working-class laborers. Nearly every student I speak to at this university discusses how, since starting college, they've learned how important it is to watch what you say online because you never know who is keeping an eye on you. They know—it is always on the tips of their tongues—that one single post can cost them everything they've worked so hard to achieve. When you are the first person in your family to go to college, the stakes are high, and their school is reminding them of this almost daily, and not just in the Career Center but also in the classroom. The university has taken on, with gusto, the task of educating its students about how to behave appropriately online.

One young man who attends this particular university, Mack, a sophomore who grew up on a farm and used to raise show pigs, speaks often of the difference between how he used to post before college and how he's trying to post now. His posts tend toward the negative—he'll post updates such as "I just had a really bad week" or "Will this end?" Mack will post when he's "really stressed" and is atypical in that he admits to using social media as a place to vent. (I heard complaints all the time from students about exactly this type of person—the one who's always negative—but it was rare that someone I interviewed confessed to engaging in this behavior, which led me to wonder, where are all these negative posts coming from?) Mack says he used to post a lot back in high school but does so much less frequently now. "One of the reasons I don't post very often is because I'm trying to keep a very professional appearance, not like a superprofessional appearance, but not anything that would discourage a future employer," he explains, then goes on to list the things he's learned not to post about since he's gotten to college. "I don't post any profanity. I won't post how someone's being terrible. I won't slander anybody or call someone out on a post. I always use proper grammar. . . . But also, if I'm going to respond, I think about what I'm going to say before I say it." Mack's university is trying to teach everyone to stop putting up the "typical party post," the photograph of underage kids who are "hammered." This is a "professionalism" issue, they've told everyone. You don't ever want to document illegal activity.

Since coming to college, Mack has begun to see a difference between the kinds of things he and his peers on campus post, and the kinds of things people from home post—something I hear frequently on this campus. "I won't be putting [up] just silly stuff, and sometimes I'll post or repost something that struck me as inspirational, but I guess my image does matter to me in the professional way, just because I don't want people to think I'm white trash," he says. "There's a lot of that in the posts of the people that are connected to me from home, just because that's the kind of demographic that we're in. There's a lot of profanity, and drama, there's so much drama. It's just bad grammar, or just blatant laziness, and texting lingo, and just, it doesn't look educated, or even a high school level of education. I try to portray more that I'm a college student and I'm intelligent enough to post well."

Another senior at Mack's university, Nikki, tells a similar story. "I used to just do stuff [online] for attention," she says. "But now that I'm older, you know, I watch what I say. I watch what I post online, due to the fact that I know that employers will be looking at your profiles now. So I make sure I don't curse online or anything like that, because I want to be hireable and able to get a job someday." Nikki won't ever post anything bad about anyone else, either. She rattles off nearly the same social media no-nos as Mack but adds to that list the importance of not posting anything "emotional," which, she explains, means anything negative, like if you've had a bad day or you're feeling angry.

Nikki has learned some of these lessons because she is a resident assistant.

"Now, working in residential life, you know that they can look us up and they're like, if we see something crazy, we can possibly get in trouble for [it]," she explains. She says it is part of her job to be Facebook friends with certain people, like the administrators who hired her, so they can monitor what their staff members are doing and saying online. Because of this, Nikki has become hyperaware of what she posts and its potential repercussions—she needs her job and doesn't want to lose it. "I'm definitely mindful for that, and I try not to post emotions when I'm feeling bad," she says. "I don't want to say anything that I cannot take back."

So what does Nikki feel she's allowed to post?

"I will post positive things that are going on in my life," she says. "Stuff like, how I'm blessed. You know, stuff that's positive. I just try to post positive things, not negative things." "Positive things" are basically all Nikki thinks she can put up online, and she returns quickly to what she *must not* post. "I won't post drinks," Nikki goes on. "You won't catch me in a bar having a drink in my hand or, you know, with illegal substances. Intimate stuff with a man—I won't post [any] of that." "I would not post sexual-type pictures. You would not catch me doing [any] of that." I heard a lot from students about "the bikini picture." The consensus is that you are not supposed to post bikini pictures because they reflect negatively on you. Words like "trashy" and "slutty" are thrown around when students mention these photos. Yet many young women at this particular institution had posted those dreaded bikini photos at some point in their pasts, and only recently learned that they had to come down to avoid ruining the chance of getting a job in the future.

Sheena used to get into fights on Facebook before she came to college and learned better. She spoke of instances in which she "was kind of angry" and "lashed out" at someone. She'd put her "business out there for everybody to see." When she was younger, she "was really vulgar and inappropriate and childish and cursed and used a lot of slang." All evidence of this has since been deleted. "Because of the schooling that I've been doing," Sheena explains. "I don't want to put that [on social media] because I don't want them to judge me or feel that I'm not appropriate to do the job that I want to do." Sheena has her university to thank for helping her develop an online filter. "I think [my social media presence] changed more when I started going to school and started to realize that I needed to be more professional and not so high school, middle school." One thing that's changed dramatically is how often Sheena posts—very infrequently now. And when she does, it's to post pictures of her nieces and nephews participating in activities such as going to an art museum or trick-or-treating.

Like Mack, Sheena notices a big difference between the kinds of things her peers at school post and the kinds of things posted by people from home who are not attending college. "There's a lot of sex," Sheena tells me. "Bathroom selfie shots, the bra, the bathing suits. . . . Just, have a little bit of class when you [take selfies]. Don't be drinking or doing drugs or just posing vulgarly. I have a lot of friends who I've deleted here recently because of it, because of nasty pictures. Or dudes even. Some of my guy friends on Facebook like to download girls shaking their butts or purring or [performing] sexual acts and I'm just like, I don't really want that on my page, so I'm going to delete you."

Sheena is always thinking about her online image. "I mean, *now* I do," she says. "Before I didn't. I didn't even care! I was posting pictures at the club and everything else. But now I am concerned with it because, like I said, I have professors who are on Facebook, people I intern for on Facebook, and I don't want them to see me in a state where I'm not comfortable, or if I were drunk or hung over and someone took a picture of me. I wouldn't want that on Facebook because I wouldn't want them to see that part of me." Sheena gives credit to her professors for helping to change her ways. They have explained how important it is to be aware of your online image, especially for employment reasons, and their advice has been very specific. "Other people in the professional field have Facebook, too, so if you wouldn't want them to see certain things on Facebook, then I would advise you to don't put it on Facebook," Sheena says she's been told. Her professors have also advised their students "to create a new account that you are more willing to let professional people see." Sheena has become very cognizant of anything associated with her real name. "My Facebook is my name, my *government* name, and some people create Facebooks with 'Weird Cupcake,' 'Sweetie Pie,' stuff like that. So I kind of thought, well, okay, if I'm going to have a friend social Facebook, then that would be appropriate, but if I was going to have a more *professional* Facebook, then I would have to use my real name. And since my real name is already being used, I kind of want to filter who's on my page and what they're posting and stuff like that."

Sheena chose the professional over the social, and she has been trying to teach the same values she's learned at her public university to her nieces and nephews, too, but she worries that they're "not really getting it." Then again, she says, "I didn't get it either."

Teaching college students not to document illegal activity is certainly worthwhile, as is talking to them about how their online behavior can affect their professional futures. This has become common knowledge for privileged high school students, so providing it to less-privileged students on campus is a matter of better late than never because employers, not to mention other gatekeepers, *are* watching.

College students who learned these lessons more recently tend to feel proud of their reformed behavior, but at the same time, they also display a sense of superiority in relation to the "people back home" who don't know any better. By comparing themselves with their high school friends, these students can see the gap that college has created between them and their erstwhile classmates, and like to point this gap out. They are learning practical skills about how to get ahead, and they feel fortunate about this in ways that students at the other participating universities take for granted or find exhausting, since they've long ago started filtering themselves online and have begun to resent the need to do so.

But what about those "friends back home" who aren't college-bound and who aren't getting this lesson? With social media infiltrating our social and professional lives more and more each day, how will this affect the job prospects of those young adults who never go to college? Will lessons around how to professionalize one's social media accounts ever reach them? Or will the gap between the college graduates and everyone else simply grow even wider?

Then, of course, we should also wonder: Does the professionalization of social media defeat its entire purpose? Are these platforms really "social" anymore? Is it healthy for young adults to gear everything they put up online toward their eventual employers, especially when so much of their lives are lived online? And what are we (and they) losing when this space that was created for social connection and self-expression is no longer a good place for exactly that—at least not when it's attached to a person's real name?

WHERE KNOWLEDGE IS POWER

There is only one university where I consistently met students who are politically and socially active online, and who think of social media as a tool for speaking freely. They don't seem to think that being opinionated or politically active on social media will affect them or their futures—or at least not in a way egregious enough to dissuade them from posting. It is the most academically prestigious university—by a significant margin—out of all the institutions I visited. The young women and men I interviewed there are some of the top students in the United States, the ones who receive perfect scores on their SATs and are so academically engaged that, at times, it can be detrimental to their social lives. Their school is at the opposite end of the ratings and economic spectrum from the public university where the administration and faculty were working so hard to undo the damage that unfiltered, uncensored social media profiles might cause its students.

It's also true that this prestigious school boasts just as many anonymous college Twitter feeds and college Facebook groups—as well as a very active Yik Yak—as all of the others, spaces where people confess the good, the bad, and the ugly. I met a number of students at this university who struggle with the usual hang-ups about wanting to get "likes" and trying to appear happy, even if they aren't. In many ways, these students seem a lot like those I met elsewhere, but they also seem to feel freer to be honest about all sorts of views and social involvements.

Bo, for example, feels it is important—almost a duty—to express his political views online. "I use Facebook a lot, and everything I put on Facebook is either a joke or something funny, or something political," he tells me. "I feel like it's some combination of actually trying to convince people and just expressing what I think." Sometimes Bo is disappointed when he posts something political and it doesn't get as much of a response as he thought it would—though he does dread, a bit, the thought of getting a really angry response. "I just try to say what I think and then if it's too controversial, tone it down, but then not very much either," he says. "And then engage with people when they say stuff." Bo's posts about Israel have sparked the most heated reactions so far, and he both loves this and feels stressed out by it. But he still considers it important to express himself openly online.

The other, more prominent trend I notice at this school is that several of the students seem completely disengaged from social media (three out of the fourteen students I interviewed have no accounts and no interest whatsoever—they live as though social media doesn't exist). And if they are engaged, their ability to maintain a certain objective distance is pronounced. In fact, they spend a good deal of time thinking critically about their lives on social media—they can't seem to help themselves. If they care about "likes," they are philosophical and self-aware about pondering what it means about our culture and their generation.

May—a fairly active user of Facebook—marvels at the ways in which we are able to create "new identities" through social media and then simply decide to "discard them." This is both good and bad, in her opinion, and her capacity to analyze what this means on a broader scale is fascinating. But her ability to think critically about social media and her own role in it also seems to give her confidence about her relationship to it; it gives her a greater sense of control, one that is markedly different from what I've seen elsewhere.

Then I meet Lin, who speaks at length about her many "metafeelings" about social media, her struggles to not get too sucked in to caring about "likes," or take too many selfies, or spend too much of the day comparing herself to others. Lin is engaged on social media, but also can analyze her engagement and that of her peers at such a high level that she seems to feel protected by her ability to critique it. There are also many other students who speak of the ways they've subverted the rules of each platform or simply sidestepped the drama altogether because it does not serve their intellectual purposes in any useful way.

I interviewed students across all thirteen institutions capable of thinking critically about social media to varying degrees—it's a key reason the interviews turned out to be so fascinating. But I also got the sense from many of these same students that their first opportunity—their first *invitation*—to reflect critically on the role of social media in their lives came during that very same interview. Prior to meeting with me, they'd never really taken the time or had a forum in which to do this. They are capable when asked, but no one had asked before. Many of them seemed to have "eureka"-type moments in the interview room, and lots of students commented after the interview was over that they'd never thought about social media in those ways before, or realized they held those opinions or feelings until they found themselves speaking about them.

But at this highly prestigious institution, students are used to engaging critically with everything. They don't need to be invited—they just do it. And knowledge—in the form of critical analysis—quite literally seems to translate into power. Being able to think clearly about social media, believing that they have the intellectual skills to best Mark Zuckerberg at his own game and understand some of the more manipulative ways that social media infiltrates our lives and relationships, gives them a healthier, more empowered relationship to it. Plus, their academics seem far more enticing and central to their lives than anything that can happen on social media; and this, too, provides them a sort of protective shield that I did not see among the rest of the interviewees—at least not so consistently.

It made me wonder: Why is this group of students so different? Why this particular school and none of the others? Is it really due to the fact that the academic and intellectual gifts among these students are so outstanding that they provide a safety net of sorts for students' future job prospects? Is it, in part, because these students aren't as concerned about their social lives as the students I met at the other schools, since their studies took precedence above all else? Is it simply because they attend such a prestigious institution that they feel more protected by their greater access to postcollege networks that will bring them professional advantages?

I asked every student I interviewed whether, during their time at college, any of their professors had discussed social media and how it's affecting our lives and our world. I'm not talking about the "professionalism talks," but about academic discussions of social media. It was very rare for a student to answer this question by saying yes. Occasionally, I found a student majoring in journalism who'd learned about social media in a classroom setting, or a student who'd taken a class that expressly engaged social media or media in general. But most students simply said no, they'd never had this opportunity; or if they said yes, they followed this by explaining that many professors forbid smartphone use in class—but the simple ban announced at the beginning of the semester was the only time the subject came up. The other affirmative responses almost always arose in the context of a Career Center seminar on the importance of Facebook Cleanups and G-rated social media profiles.

But critical analysis about the role of social media and new technologies in our lives and world, and how it affects the way we construct our identities and relate to others? These subjects were nearly nonexistent in classroom conversation in these college students' experience—despite how much social media has taken over their lives (and everyone else's). Students are being taught to "professionalize" accounts attached to their names, but they are not being challenged to think about social media during their studies. This was true for the young people at this prestigious institution as well, but these students had come to college with such a high level of intellectual engagement that it seemed natural for them to apply those skills to social media too. It's just what they do.

Yet, if knowledge is indeed power, we must ask ourselves if a Career Center presentation about how to maintain a spotless social media presence is enough to prepare our students and other young adults to effectively handle social media. Or even whether promoting the professionalization of social media could be detrimental because it supports the idea that appearances are everything, and reality should be hidden. We are giving college students a kind of education around social media, but it rarely crosses the threshold of the classroom, the place where—at least in theory—they should be learning to think critically about the world.

There is both an intellectual and an emotional benefit to such engagement. While it may not prevent young adults from struggling with social media, the capacity to process what's going on intellectually can help them feel as if they are the ones steering the ship, navigating social media, and that they are not merely at its mercy.

There is another side to this professionalization for students, too: the feeling that they need not only to avoid mistakes but also to actively market themselves. And that has its own complications, as we shall see.

3

MY NAME IS MY BRAND AND MY BRAND IS HAPPINESS!

You are a brand, and social media is the platform on which we project our brand to the world.

Annika, sophomore, private-secular university

Even I must admit that when one creates an online profile attached to their real-world identity, they create a brand for themselves. This brand, just like Kellogg's or Ford or Nike or whomever you please, there is an image to uphold and if it gets stained, you better have some strong bleach.

Nancy, first-year, Catholic college

CHERESE: A NONSTOP "PUBLIC APPEARANCE"

The moment Cherese walks into the interview room, the atmosphere changes. She's a woman in a hurry, with things to do. She has promised to be here, so she'll fulfill her commitment, but she's not going to stay any longer than necessary. Short and round-faced, Cherese attends a small Christian college in the Midwest. I try, as usual, to start our conversation with small talk, but Cherese is having none of it. She's all business, so off we go, a bit awkwardly, Cherese's deep voice filling the room with abrupt answers. I quickly learn that she's a junior from Chicago and a serious debater. She's also very involved in the missionary Baptist church her grandfather founded. Her aunt is the pastor and her father the associate pastor.

For the entire first part of our conversation, she sits poised on the very edge of her seat, as though ready to make a run for it at any moment. She sighs and huffs as I go through each of the questions. Then, things finally turn to the subject at hand, social media, specifically Facebook, and Cherese's demeanor changes. She edges back into the chair a bit and looks directly into my eyes.

Cherese suddenly has things to say. A *lot* of things. Things she seems ready to get off her chest, that maybe she's wanted to say for a while but has only now found the right forum (or the anonymity). She launches quickly into a litany of complaints about how you really have to watch what you post online because of what certain people will think, and how she spends a lot of her time worrying about who she can allow to see which posts and who shouldn't see any. Cherese restricts certain posts to certain groups. This prompts me to press her about what criteria she uses in deciding what to show to whom, which is when she tells me something that makes my jaw drop.

"I have seventeen different [Facebook] groups," she says.

"Seventeen?" I repeat, stunned.

"So I have a church group," she begins. "And then I have the greater Chicagoland area, and then I have people from churches that we go around to, some that are in the DC and West Virginia area, but then I have a different group if they don't believe that Baptist people should be doing this, or because they just don't believe that's [within] the doctrine of faith. . . . I have three different [college] friend groups. So, if I was just writing something to the general [college] population, then it just might be something about a class or something about a major, or I use it for advertising something from the office or something else that's happening on campus. But I have my multicultural [college] friends list, then I'll post something . . . . like, 'Student activists taking control at another college or university' or something that they will feel empowered from looking at." Cherese goes on to tell me that she blocks her mother from most groups (though they're friends on Facebook), and she lets her father see just about everything she posts because he's "more accepting."

"Is it difficult to juggle seventeen groups?" I ask.

"Yes," Cherese agrees. "It just becomes like, 'Well, if I get ready to post this, then what am I getting ready to say to *this* person, and what am I going to say to *that* person, because if I say *this* to *that* person, then they may be offended or they may have other questions and because, then you have to change [the post] and see who you're trying to say something to. I won't say I'm afraid, but I'm just very conscious about what I get ready to post to certain groups of people. . . . I'm an activist, so I post something about gay rights and then [someone goes], 'Oh my, we should talk to you about your sexuality. Have you been questioning your sexuality?' So then, well, I'll put them in this group on Facebook so they won't see when I post something else."

I ask Cherese if her fear of offending someone is related to her religious background. She tells me that is part of it, but it's more that she likes to post about "revolutionary things," by which she means racism, feminism, and sexual orientation—things that provoke strong responses from people, and not always the kinds of responses she's seeking. When one of her Facebook friends seems unable to handle a certain comment, she relegates that person to one of the many groups that she doesn't allow to see those sorts of opinions. Actually, very few people get to see those provocative, opinionated posts Cherese cares about so much.

"Why don't you just unfriend those people instead?" I ask, thinking that this would simplify everything.

"Because that'll cause *another* problem," she says. "Well, [and] because then I would have, like, zero people because I would delete everybody."

Cherese truly believes that her only option is to engage in this continual, frenzied dance of multitasking, managing seventeen groups, trying to please (or at least not upset) a diverse array of "friends." She's gotten herself into this situation by accepting friend requests from people she obviously finds difficult to please (or just plain difficult), but she doesn't feel able to liberate herself from them. It's obvious, too, that Cherese wants at least *something* of an audience. She wants to be able to share her opinions and, ideally, to be heard. She wouldn't like having "zero people" on Facebook, even though she implies that it would be easier on her. By creating these groups, Cherese has found a way both to maintain a wide audience and to be heard only by those she believes will agree with her opinions. Like so many students, she's conflict-avoidant. Cherese is a young woman who knows she can be provocative, yet she's figured out how to use social media so that she expresses her provocative views only to those who won't find them provocative.

And while just about all college students are hyperaware of the future employers who might (and surely will, at some point) be watching them, some, like Cherese, worry about the image they project on social media in a much broader way. Cherese's interest in social media is still, well, *social*. She's incredibly image-conscious, but the juggling of her many groups seems rooted in this desire for—and acute awareness of—an audience from which she is seeking approval (or, at least, not disapproval). She isn't worried about "likes" as much as someone like Rob, but she wants reassurance that there are people out there seeing what she posts—the *right* people.

Social media is the new stage for performance, and as the actors in a play or musical hope for applause, many college students are learning to "play to" and actively tailor their "performances" to the crowd. They learn to manipulate the audience. And Cherese has become a good manipulator, though this doesn't exactly make her happy.

Cherese and I spent a good deal of time talking about happiness, too—how it often seems to elude her, how she doesn't really know what happiness is, how she thinks she'd probably be a lot happier if she stopped worrying about what other people think and how other people define happiness. Cherese's housemates are always telling her that she doesn't have any fun because she never goes out like they do, but going out like they do isn't her idea of fun. "I get excited if I win stuff, if I can win a prize or something," she told me. "I'm a very competitive person, but I also like winning souls. . . . Basically I'm an evangelist, so if I can bring people into the church, that's exciting, or something else exciting is to do something for a social cause, anything dealing with an 'ism,' like, to bring down stereotypes."

This is one of the reasons Cherese needs so many Facebook groups. What makes Cherese happy isn't necessarily what makes her peers happy. In fact, the things she associates with happiness are often things that upset other people. She finds satisfaction in fighting racism, for example, but some people don't want a continuous string of serious and some-times seriously depressing news on their newsfeeds. By sorting friends into separate groups, she's both protecting those people from seeing posts that might upset them and protecting herself from constantly having to reckon with how her definition of happiness is vastly different from theirs—a fact that both pains and annoys her. Cherese has learned that the best course of action is to display the *right* kinds of happiness to those groups that will appreciate that particular version of it.

Cherese has also learned that if she posts anything that is the least bit negative, everybody freaks out. She now knows how to turn a negative into a positive so she can ease everybody's mind and play into the reality that social media is about everybody feeling good, all of the time. Posting on social media, she says, is about "feeding" other people's "stereotypical sense of happiness."

"I don't just post like 'Hip, hip, hooray!' [or] 'I got a 100 on a test,'" she explains. "Basically, you pick one good thing that happened so that you're not just posting bad and depressing things. So I turn the negative into a positive. Like, yesterday wasn't such a good day at all, but my peer mentor helped. So [my post] goes, 'Thank goodness for peer mentors,' and then everyone took it as something very positive and happy."

"You just leave the 'not a good day' part out?" I ask.

"Yeah, just 'Thank goodness for peer mentors' was the exact word-ing," Cherese confirms. "So it was just, have the happy moment so that people flourish and think, 'Oh yes, she's happy now. Okay, so we won't have to say anything else.' . . . . If you post something bad or if you say, 'Oh, my day wasn't so bad' or 'My day was just, really bad today,' then ev-erybody gets to calling and becomes concerned . . . . and then it just kind of cause[s] a really bad backlash . . . . So it's like, 'Well, I'll just post: "Today was an excellent day,"' and then nobody will come to say anything."

I ask what seems like an obvious follow-up question: Why keep posting if you can't really be honest? Cherese laughs at my naiveté. If she doesn't keep posting things, she explains, then people also become con-cerned. Posting, in general, is like "a public appearance," she says. You have to pop up every once in a while to prove you are okay. Cherese refers to it as "people-pleasing." Constant people-pleasing can be tiring, she says. "If you're just going to say one happy thing, it does become exhaust-ing, overwhelming, because you're always thinking, 'Is somebody looking at this?' and it's like, if somebody's looking at it, who's looking at it? So then it's like, 'And how will it be interpreted?'"

Cherese isn't exactly sure why people need to see that you're happy all the time. "I think that they just want to see you conform," she says. "I think people believe that if you're happy, then everything is good, so you don't have to do anything. So the more they could see you happy, then—I don't think it'll make *them* happy, but it just seems like, 'Oh, she's like us.'"

Cherese is getting at more than just the need for mere conformity. In her eyes, people want the relief of not having to worry about anybody else, or being required to do something in response to another person's pain. If everyone plays along and pretends that all is well, then the "audience" gets to feel at ease, even apathetic, about everything they see. The happiness effect, in this case, is one in which viewers get to remain just that—passive viewers, scrolling through the feed and nodding their heads and never really having to engage anyone on a level that is real.

THE (HAPPY) FACE OF SUCCESS

Cherese's management of seventeen Facebook groups is unusual. But the general notion—dividing your life between what's post-worthy and what isn't, and maintaining a carefully crafted facade by posting only positive things—is typical. This division has repercussions far beyond social media. It has both emotional and social consequences for young adults, who suffer from the dissonance between what they see others post and what they feel about themselves and the realities of their lives.

"I think sometimes people think, 'Oh, I have all these pictures, and friends on Facebook, and they're all happy and successful,'" says Gina, a junior at her Midwestern Christian university. "I feel like sometimes it's kind of a wall, like, you try to take a picture and be like 'Oh, look how happy I am! Look how great my life is!' But in reality you might not feel like you're that successful."

I press Gina to say more about what she means by "successful."

"Like, you could be posting a picture from this ceremony or from graduation or something, which is extremely successful, and a huge, monumental period in your life," Gina says, referring to that all-important "highlight reel." "But then, you could just be, I don't know, putting on a happy face for a picture to show other people that you're happy."

On a basic level, this propensity to post only happy, positive photos and comments seems completely reasonable. Who wants to put up tearful, unflattering, unsmiling images of themselves for the world to see?

But the conflict or, perhaps more accurately, the *burden* students feel with regard to both kinds of posts is palpable. It requires maintenance on their part: because social media is a constant in their lives, the expectation to post those highlights and smiles is also constant. To post something less than happy or to stop posting altogether might set off alarms among the people who know and love you, as Cherese worries. Worse, if it *doesn't* set off alarms, it might confirm your fears that you are invisible to everyone else and that nobody really cares about you. Then, there is the fact of having to see everyone else's happiness and highlights all the time. Being constantly bombarded by smiles and successes when you're feeling low—seeing, as Emma described it, the best version of others versus the worst version of yourself—can be difficult to endure. The way students talked and talked and *talked* about making sure to "put on that happy face" for everybody sometimes evoked for me those maniacally smiling emoticons that give off a kind of frenzied joy—happiness with an edge, if you will. That is the self-image students are learning to project.[1]

Lucy is a junior at her Catholic university, and she's very focused on academics. She has a round face, glasses, and long brown hair, and she is quick to smile. She's also very devout.

"I choose to post happier things," Lucy informs me as we are talking about who she is online, and how she portrays herself on her social media accounts. "I got a good GPA this semester, or I'll post pictures with my family. Things that are going well in my life."

"I want people to know that I'm happy and doing well," Lucy says, emphasizing the importance of demonstrating her happiness for others. She does this many times during our interview. "I want people to respond in a good way and be excited for me. I want more encouragement or. . . . I don't really know how to describe it. I guess, it's just that positive affirmation that you get from posting something that I enjoy."

Lucy likes to post the exciting things that happen in her life because it pleases her to be congratulated by others. She posts so people can know what's going on with her, but she also thinks "it has something to do with being successful." Lucy has learned that happy updates get her that positive affirmation she so desires, whereas the not-so-happy ones often result in a deafening silence—something to be avoided at all costs. Lucy is different from Cherese in this regard. Cherese agonized about not wanting to make other people worry. Keeping her posts positive made it so none of her "audience" would feel a need to do anything—at least not anything really taxing—to help her. All they needed to do was click the "like" button. But Lucy doesn't post anything negative because she's afraid that even when others see that she's having a difficult time, they won't care enough to say anything, and she doesn't want to make anyone uncomfortable either.

"I don't think I would post something that is not going well, or I'm struggling with," Lucy explains with a long sigh. "You kind of put on a happy facade, kind of just share things that are *good* important in life, not *bad* important. . . . So I think I'm pretty happy online. Maybe it's not representative exactly of who I am in life because I *do* struggle with things and I'm not happy all the time." But Lucy knows from experience that such thoughts are not appropriate for Facebook. Seeing depressing comments from others on social media makes Lucy uncomfortable. "You don't really know how to respond to it," she says. "It's okay to say that you got a good GPA because it's something that's not going to, you know, reflect bad on your character."

Character is something Lucy considers often when it comes to who she is online, and what she and others post about themselves—not character in the moral sense, as Aamir thinks about it, but more like character in the fictional sense. In fact, Lucy thinks about "creating that character" nearly constantly. "Not that [this character] is completely different from who I am in life," Lucy tells me. But she tries to post only about happy things or things that are exciting, she adds, in order to create that "character." Lucy's concern with her "online character" reminds me of that very particular "sorority image" that Emma's sisters required of her, and how Emma's sorority house went so far as to monitor its members to make sure they were putting forth the image the sorority wanted to show to the public.

I press Lucy to say more about what she means by "creating a character" on social media. "I think I am true to who I am on Facebook, but I just leave stuff out," she explains. "I wouldn't post insignificant things, or things that are, you know, not happy. I don't think I'm being wholly truthful with what I post. I mean, what I post *is* true.... I don't think I edit those posts that are true to me, so those things that are online are true to me, I just leave certain things out that might make it more true to my identity and who I am." Lucy stumbles over her words as she tries to explain how her decision to post only happy things still somehow represents who she is, yet at the same time is not really a true representation of who she is because she's not always happy—even though it appears that way if you visit her social media profiles.

It's hard for Lucy to know what to make of the differing versions of herself. On the one hand, Lucy should be thinking about her character and what might reflect badly, or positively, upon it. This is a mature attitude to have about one's online activity. Lucy thinks that focusing on one's online image and creating a certain character on social media can be a healthy thing. It gives her the means to explore different ways that people might perceive her and different ways that she perceives herself. But this can be negative, as well, she says, "because you can spend all your time focusing on how your character is perceived by other people and [wonder constantly] if you are doing the right thing." Sometimes college students like Lucy become so concerned about not posting anything that could reflect badly on their character that they develop a crippling fear of failure and a concern about "appearing happy" so extreme that it can eventually sound almost pathological.

In addition to character, Lucy thinks a lot about online "image." She doesn't like that she thinks about this at all—it seems somehow that "you're not being fully truthful," she says, to be so concerned about it. Yet, posting that you "got a 4.0 boosts your image," Lucy says. "So things that I would post would definitely increase my image.... I don't want to post something that might be inappropriate or not positive because I guess I do want to uphold that image that I have."

In the survey, I asked students if they feel the need to be curators of a particular image. Seventy percent answered yes, and an additional 16 percent answered yes, to a degree.[2] Several wrote that online image is important to think about because you have to consider the "audience" you are reaching; one of these students went so far as to say that "you pick a target audience (sometimes unconsciously) and you curate. You form your image and identity around whomever you're trying to please."

Some students didn't think this was a bad thing. The fact that we can curate an online image "is kind of like putting on a mask," one student wrote. Another felt that "it's a game, it's an art," which can be fun and harmless. It "allows users to create an imagined identity," a third student said. The problem for this last student is that these imagined identities "are often false" and, for better or worse, "these images are important, because they are shaping generations to have unrealistic standards for their life experiences."

There was also a student who described the cultivation of image online as an opportunity to give "one version of yourself permanence and authority," and another who argued that this image-consciousness "allows for creating those [positive] impressions efficiently." Gone are the days when we floundered around the lunchroom, puzzling about the successes of the popular kids, or the successes of anyone for that matter. Now young adults have the "authority" to attack such problems "efficiently."

Overall, students' responses revealed that they feel a lot of frustration and stress in relation to their online images. Many expressed a wish that we didn't need to worry about online image at all.

Asked whether people today spend a good part of their time considering their online image, one student wrote: "Very true. Our culture today is forming around 'being happy,' and although that is good, I feel people actively neglect the fact that life has ups AND downs. Therefore, social media is used only to highlight the ups of life, while the downs are more often internalized behind the walls of our bedrooms, homes and personal lives. Although I do not feel that social media is a place to air negativity, I think it is okay to not be 100 percent happy all the time, and social media promotes the latter to the extreme." Yet another student wrote: "People want to see others as happy, and people are easily bothered by someone who confesses that they aren't happy or aren't what everyone wants them to be."

Lucy is certainly showing a piece of herself online, and one that is connected to a certain version of reality, but many students feel like they're required to be fake, and nearly all students complain that what people see of them online is often pretty far from the truth of who they really are. And the dissonance some students feel between what they put online and how they truly see themselves can become extreme.

WE ARE OUR PROFILES: THE "REPUTATION SELF" AND "THE MANICURIST"

College students are acutely aware they have an image not only to protect but to *create*. Doing so is important because a vast potential audience is out there waiting—waiting to be entertained but also to pounce when they see something they don't like. If someone trips up and says the wrong thing, they might never live it down. However, students also feel like they have a lot of control over their profiles, that they must *assert* control over them. It's essential to the construction, maintenance, and promotion of a certain kind of "self." Most college students are highly aware of how to navigate the new public construction of self on social media, and many of them are incredibly savvy about negotiating exactly how they are seen by others. They worry about their image like celebrities and politicians with teams of handlers might, and these negotiations play into this pressure to appear happy at all times.

They are acutely aware they have reputations to protect—that they "are" their reputations.

Take Brandy, who spoke at length about the "multiple" kinds of "selves" we have today, and in particular about our "reputation selves,"—a phrase Brandy used in her effort to describe the public dimension of her online persona, which is multifaceted. "I feel like [the reputation self] is another 'self' that now, considering social media, is bigger than ever because there's who you are in how other people see you," Brandy says. She thinks that everyone has multiple selves: the physical, the spiritual, and now the reputational. "That's a third version of yourself because you basically decide what you put out there and that's the way you want people to see you," Brandy explains. But she sees the "reputation self" as something new and particular to social media. "The way you want people to see you [online] isn't a true reflection of yourself but that's still a *version* of yourself," she adds.

I ask Brandy why she thinks the online version of the self isn't a "true" version. "Because it's not real," she answers simply. And here we come back around—as all my interviews seem to do—to the issue of appearing happy. "It's not like you're going to post about all the terrible things that you've gone through." Brandy thinks that someone showing an authentic self is "something rare that you never see anymore." When I ask her why, she responds, "Because I feel now, more than ever, especially with social media, you're just given the term 'profile' and you *are* a profile." This isn't right, according to Brandy. People shouldn't be just profiles because profiles can't encompass all that we are. Brandy thinks the existence of social media profiles makes all of us try to fit ourselves into "boxes," something she doesn't like and doesn't feel capable of doing herself; the "boxes" social media profiles allow are simply too general, like "partier" or "Jersey girl." "You can't really build a person in a checkbox kind of form," Brandy says. "Now, since people are more exposed on social media, it makes them more anxious than ever that people are going to judge them, just because that's what people will do."

The way others perceive us is extremely important because of social media, in Brandy's opinion, and this is due "to the fact that people are now able to know, if you want people to know, everything happening in your life on a second-by-second basis." It bothers Brandy that what she sees her friends posting online is just a facade. When she goes to their pages, she thinks to herself, "This isn't the real you." Instead, "People have pressure now, more than ever, to project an image that everything's peachy and wonderful in their life," she explains. "It's kind of like how everybody says with their high school reunion, they want to go back and show off how great their life is. It's like that now, but you don't have to wait for your ten-year reunion. It's like that *every day.*"

Brandy isn't immune to this pressure. When I ask if she ever gets caught up in this herself, she exclaims with a big laugh, "Absolutely!" But then she qualifies her answer, saying, "But it's not with the intent to make, to project that I'm better than somebody else." Sometimes Brandy feels guilty boasting, but everyone around her faces the same pressures and, as she explains, "You can't help but get caught up in it after a while."

The pressure is difficult to resist. Image, as the old slogan goes, is everything.

Hannah, the same young woman who can't help comparing herself to her sorority girl roommate, really dislikes social media. It feels like an obligation and a not altogether pleasant one. "It's time-consuming," she says. "It's *so* time-consuming. . . . I definitely feel like [social media] is a job, like, I have to curate everything I put on." Hannah uses one of those words—"curate"—I heard often from students. But then Hannah offers up a new metaphor: the "manicure."

"I think people want to manicure their lives and put up a facade that is what they want people to see instead of what their actual lives are," she tells me. "I think we *think* that they're successful, but I don't think that they necessarily are. For example, I was talking about this in my family studies class the other day, when people have kids, and then they only post the things that are good about having a kid and they don't talk about how the kid just threw up on them five minutes before they posted the picture. [Other people] think, 'Oh, they're a successful parent. They're a happy family because these pictures all look so good.' But then, in reality, the family can totally be failing. So we *think* that they're successful, but that's because they're manicuring their presence." Hannah has a lot to say on the subject. "This is definitely something I've thought about before," she explains. "It *is* like a manicuring! It's going to look perfect and . . . . it's a facade. You're making [your profile] *exactly* what you want it to be instead of being what it actually is."

I ask Hannah if she "manicures" her profiles. "If someone puts a picture up of me online that I do not like, I'm like, 'Please take that down. *Please* take that down, I do *not* appreciate that,'" she says. "But if someone puts up a good picture, I'm like, 'Oh, leave it up, yeah.'" Hannah sighs heavily. "I think that's why people like [social media]. It's because they can manipulate it, and I'm definitely subject to that too. I mean, I definitely think about it, but not in the same way. I'm thinking more about if people are going to judge me if I put up something too personal."

Brandy and Hannah are talking about something different from mere professional concerns—though they have those, too. They have an acute sense that the construction of the self online is a kind of performance. There have always been many dimensions of the self, and people have always had both public and private selves. But the online world has taken the construction of the public self to new extremes, requiring this generation (and anyone who operates on social media, really) to contemplate "self-image" in a far more heightened way.

THE THREE CS: CRAFT, CULTIVATE, CURATE (AND A "MARKETING CAMPAIGN FOR ME")

To be human is to be social, and to be social is to have an audience. I live in New York City, and each time I ride the subway I have an audience— even if I'm not paying attention to it and my fellow subway riders are not paying attention to me. The same goes for those who grew up before the existence of social media. We'd walk the halls at school and sit in the lunchroom with our peers. Whether we realized it or not, we had an audience, and we played to it, too. There were informal rules about dress; there were attitudes and interests that divided us into groups; there were popular kids and unpopular kids, and you could move up and down the social ladder depending on how a particular audience perceived you— or didn't. There were groups of kids who excelled at rising socially—the best and shrewdest manipulators among us—and there were those of us who bumbled along trying to figure out how everything worked, trying and failing, trying again and occasionally succeeding. But no one, not even the queen bees, explicitly spoke about audience and their particular, personal "brand," which they needed to "protect." We were utterly naive about this or, perhaps, blissfully unaware.

But, unless we become hermits, we live our lives amid a series of "publics." There is an innocence to not having this responsibility openly and obviously sitting upon our shoulders, a freedom from actively knowing and having to manipulate and play to a variety of audiences. Until recently, it was the kind of awareness and knowledge that only a certain breed of adults—primarily academics, marketers, and advertisers—had to endure.

But today, many young adults are aware they have a series of "publics," and they are growing up learning how to actively navigate and even manipulate their audiences to achieve certain outcomes. Because of social media, we are becoming master manipulators, constant performers, and no one is better at these endeavors than young adults, because they are learning earlier and earlier that these skills are central to success, either social or professional. In *Reclaiming Conversation: The Power of Talk in a Digital Age*, Sherry Turkle writes extensively about how social media and our technological devices have allowed us to engage in a near-constant "editing" of speech as we text and email and chat, which can make in-person conversation not only daunting but, for some, so anxiety-inducing that they avoid it altogether.[3] After speaking with and surveying so many college students, I worry that they are not only learning to edit speech but also learning to "edit" their selves for "publication" online.

Perhaps the most interesting aspect of all my conversations on this topic is the language students use to discuss how they've learned to operate online. They often default to business jargon. The idea that we can "advertise ourselves" through social media is common, and one student went so far as to say that the image one cultivates on social media "can even be used to create a business of being 'yourself' online."

The production of self is itself a business enterprise.[4]

Take Ming. Like Brandy, she is acutely aware of the importance of online image, but she resorts to business jargon to explain it. "It's kind of like a marketing campaign for a company," Ming explains. "You want to show off either the best parts of yourself, or you want to show off, maybe somebody's an activist, and this is the part that they want the world to see of them. It's not a core part of you, but it's maybe an extension of yourself. The side that you like to show the world publicly." Social media is the perfect framework for launching a particular self-image or for showcasing an especially promising aspect of a person's life. Businesses have specialties—that's how they succeed—and one's online presence also requires focus, promotion, and specialization for it to "succeed." Even when someone doesn't explicitly intend to do this, it happens anyway. "I don't think, for the most part, people intentionally think of it as, 'Oh, this is a marketing campaign for me, and I want people to see all the best parts,'" Ming says. "[Social media] just naturally allows us to show everybody, 'Hey, these are the cool parts of my life,' or 'These are the interesting things that I'm doing,' while you can relatively easily omit the things that maybe you don't want other people to see."

This is another reason Ming uses business terms. She thinks that "certain industries" really pay attention to social media, and whether you have a "Twitter presence," or "a blog that you keep that you feel represents your work or your social media identity."

Another student, John, speaks of social media as a means for advertising the self. "I think [social media] is a good way to market," he says. "I think you can market yourself through it. I think you can market your ideas through it. I think there's a lot of good aspects like that. Marketing and staying in touch."

When I ask John if he "markets himself," he answers, "I guess in a way, yes. Because you, *I*, try to show myself in a positive light. I don't try to be too negative or post bad, inappropriate things, and I post about internships that I get or these good things that I'm trying to do, like the job I want to get." For John, social media is aspirational. He shows only his good side and references the things he wants in life. "In a way *everyone* is marketing themselves because in the end, someone, your *boss* eventually, is going to see it, and he's going to like or *not* like what he sees." John goes on to discuss his many criteria for posting, and as he's listing these rules, he starts to laugh. "Yeah, I definitely make a conscious effort to only post stuff that is okay."

Then there is the first student—though not the last by any means—who describes herself as a "brand."

Fara is petite, with a delicate frame and long, silky black hair. She was born in Jakarta, Indonesia, and she is so soft-spoken during our interview that I have to repeatedly ask her to talk louder so I can hear what she's saying. A senior at her public university, she, too, defaults to business terms to describe her life online, but she does so in a far more intimate way than John or Ming. Fara is more like Cherese in the sense that her social media existence is central to her self-understanding. She is extremely shy in person, but on social media, she's *very* outgoing. And she's very serious about her profiles. She sees them as a public extension of herself, which means she must cultivate them carefully. Online image is everything for Fara, but unlike Brandy, she doesn't think of this as "untrue" or false. But she does speak about reputation.

"I think of myself, like, my name, as a brand," Fara says. "So I like to stay active on my social media platforms, but I choose, I *select* when I share. . . . I have a reputation and I need to protect it. So I don't share things that are private, things that are going on in my romantic relationships. I'm very selective, I'm a curator." As with many students, selectivity is important—both the frequency of posts and what you choose to post. You never want to share too much, and you only want to share exactly the right things. "If I think I'm oversharing, I would delete that post," Fara explains. "It's a matter of being careful about what you share, thinking about who's reading, who's looking at this post, because you never know if your professor looks you up and you're complaining about her class."

Fara's notion that "her name is a brand" goes far beyond worries about professors or future employers. "Anything that is associated with your name, [future employers, everyone really] is going to find out, so that's why I think of my name as a brand. I'm *very* careful about what I do online and how my online presence is."

I ask Fara what sort of "brand" people encounter if they search for her. "I tend to share what I'm doing in school, and it's not provocative," she explains. "It shows that I'm a responsible person and I do have connections." Fara also has a heightened sense of audience—that she has one and that, just like a celebrity or a famous actor or writer, Fara must keep her public in mind when she posts. Like Ming, Fara worries about how many people you can reach through social media, and she wants to grow her audience as much as possible. She tries to expand the number of people that read her profile by posting only during "high-traffic times" as a way to attract attention but also to control the attention she attracts. "Again, I *expect* people to read [what I post], so I tend to post at a certain time during the day when I know the *most* number of people are online. I crave attention," Fara admits. "And I want to elicit a response from someone, so that's what motivates me [to post], knowing that somebody else is reading." When someone "favorites" her tweet or "likes" her post, it's an indication that she's managing her "brand" well and protecting her reputation. Fara spends a lot of time deciding what to post, and she goes to her friends for advice because the stakes are so high, asking them which posts they think will get the most positive feedback. "I always worry about getting negative feedback, because sometimes it ruins your day and it's discouraging. It's disheartening."

Maintaining the "name brand" that is Fara can be exhausting—so exhausting that Fara stopped posting on Instagram for a while. She was so caught up in making others believe she "has a fabulous life," that "she went here, she went there," that she had no time for anything else—though she worries that not posting there might damage her "name brand" too.

"Especially my generation, right now I'm looking at all my friends, they have personal websites where they're posting their portfolios online, and whenever we meet someone, it's like, 'Oh, you're on social media? *What* social media?' It's not 'Oh, what's your number? It's 'What's your social media handle name? Or what's your Twitter handle name?'" Fara pauses a moment before continuing. "You know what I mean? I think [social media] is very important. It is a job."

I've wondered what accounts for this language of "marketing" and "branding" of the self that so many students used during the interviews, and how much of it has to do with the seeming and widespread professionalization and commercialization of *everything*. Even colleges and universities consider themselves "brands" that must be maintained and protected, and students are regarded as consumers, so the educational institutions themselves promote such language. But there are also countless examples of bright, savvy young adults who've achieved widespread notoriety and celebrity by creating their own YouTube channels and

Instagram pages with thousands, even millions, of viewers, using social media to turn their lives and lifestyles into money-making "brands," and who are regarded as new and valuable tastemakers. Today there is such a thing as "Internet famous," which comes with the kinds of perks that used to accrue only to Hollywood celebrities. Companies looking to sell clothing and other wares to the young and moneyed see these "name brands" on social media as potential gold mines.[5]

While the students I met did not seem interested in becoming Internet famous on any widespread level, they'd certainly internalized the notion that on social media you are always marketing, advertising, promoting, and producing a public persona that can make or break you, so you'd better proceed with caution, care, and a good deal of savvy. The "production" of self for one's audience—both the immediate one and the potential future one—has become second nature.

It's just how things are.

Perhaps most revealing of all is the answer students gave to a question in the online survey where they were asked to reply yes or no to the following statement:

I'm aware that my name is a brand and I need to cultivate it carefully.

A total of 727 students responded to this question, with 79 percent of them answering yes.

Even as young adults are crafting, curating, and cultivating a particular online image for their audiences, and engaging in so much people-pleasing and self-promoting behavior, they are also aware that it is not "real" in the way that reality TV isn't "real." They understand that much of the "happiness" they see displayed all around them is produced for a particular effect on the audience.

But the expectation on them to pay attention to and craft a particular image? *That* is real. And the "branding" of the self comes at a great cost to some.

Whether the students are ambivalent about the reality they perceive around social media and image, or frustrated by it, the overwhelming feeling they have is that online image is hugely important, and if you care at all about your future, you'd better start curating your image.

Sometimes that image is metaphorical—the sum total of all your online identities—but in some cases it is literal. Which brings us to the selfie.

4

THE SELFIE GENERATION

Why Social Media is More of a "Girl Thing"

I think our generation in general is just in love with themselves.

Abby, first-year, evangelical
Christian university

I would say that the whole selfie thing is predominantly female. I actually do a lot of unfriending because of that.... It's an image thing. It's always wanting to look pretty, to be accepted.

Gray, junior,
Catholic university

Fifty years later, when people look upon this generation and this time period, I'm sure they're going to dub it "the Selfie Generation."

Tanuja, senior, evangelical
Christian university

THE GOAL OF A SELFIE

"I used to think that [selfies] were conceited," says Tanuja, a senior at a Christian university in the West. "I used to think they were very, *very* conceited." I ask her what changed. "Maybe it [hasn't] changed, but just that they're so prevalent, it's just like 'Okay, well, the world is doing it, you can't really help it.' . . . . Some selfies can be very fun, you know what I mean?"

At this point, Tanuja starts parsing out a selfie hierarchy: "I think the people who only post selfies of themselves without anybody else in them are still a little conceited," she says. "And there are definitely people and pages where it's just of their face and nobody else in it. Those are the ones I definitely still question a little bit." Tanuja is not alone in feeling that, if you are going to take a selfie, you should make sure you're with other people. I heard that from many students, though plenty of them are still disposed to pucker up for the camera when they're alone. But Tanuja pushes deeper. She begins to wonder how these supposedly self-indulgent snapshots will be seen in the future. "Fifty years later, when people look upon this generation and this time period, I'm sure they're going to dub it 'the Selfie Generation.' . . . . But when people look back on this, I hope, it's not going to be so much at how conceited these people were, it's going to be [more] like, look at the types of funny pictures that they took, and how they took them."

It's noteworthy that Tanuja worries about her generation being labeled "conceited," which speaks to the reality that her generation has already been labeled as self-obsessed on an historic level. The work of Jean Twenge in *Generation Me: Why Today's Young Americans Are More Confident, Assertive, Entitled—and More Miserable Than Ever Before* argues this, and the rise of selfies as emblematic of this narcissism and entitlement isn't helping thwart the label. Though Tanuja is aware of her generation's reputation (as are many other students I interviewed), she would prefer everyone to be remembered as fun and creative. There is a defeatist quality to Tanuja's sense of the negative label wider society has applied to her generation—she knows that she can explain through our interview that selfies (and all that goes along with them) are simply her generation's way of playing around, but has little faith that those who are in a position to judge will take her claim seriously (because the judgment has already been handed down).

But selfie culture also seems closely tied to the pressure students feel to appear happy—and popular—online. Selfies are usually about showing your best face, your prettiest look, your most amazing outfit, the day when your hair looks just perfect. They are literally all about appearance but also about *with whom* you are appearing. You don't want all of your selfies to be solo, lest those perusing your social media profiles think you are completely self-centered. For the selfie aficionado, it's a delicate balance.

Elise, a sophomore at a Catholic university, shares Tanuja's feelings about selfies. "It's funny, I used to hate them," she tells me. But then, like Tanuja, she began to distinguish between different kinds of selfies: what makes a selfie good or bad is in the intention of the person taking it. A selfie should not be for gratification of your own ego.

"I used to think the goal of a selfie was to get people to 'like' a picture, and if you got 'likes,' then that meant that a girl was really, really pretty," Elise explains. "And if another selfie hardly got any 'likes,' then that meant nobody thought the girl was pretty. And I hated that whole concept. I thought that was awful." Elise seems not to even ponder the possibility that guys might take selfies, or that anyone would associate taking selfies as a guy thing. At least initially, she understood them as exclusively for the purpose of other people evaluating a girl or woman's appearance, and this disgusted her.

But then Elise had a change of heart about what a selfie can do for its taker in a positive way. A selfie can be *good for you* if you go into it with the right attitude—and attire. To Elise, "It all depends honestly, [on] what you're wearing. If you're wearing a low-cut shirt or you do your makeup really, really elaborate, then you're calling attention to yourself. And I think that's your sole purpose for putting that photo up." If your intention is to get "likes," the selfie is impure. Selfies should *not* be about getting "likes," and you should not wear a certain outfit or do your makeup a certain way solely for the purpose of taking a selfie. Yet this is a fine distinction that is difficult to discern from an image on a screen. "But, you know, if you just have a pretty day, like all girls have," Elise says, "and you take a picture, ...I think that is a perfectly acceptable selfie, and I'm guilty of that too." If the selfie is incidental to the outfit, the hairstyle, the makeup, then it's acceptable. For Elise, selfies offer a way to capture a moment when you are feeling attractive and good about yourself—it's about the taker of the selfie making this judgment and acting accordingly by snapping the picture. The caption plays a role as well. As Elise tells me, "It all depends on the purpose of putting that picture up and the title. If the title's something like, you know, 'Look at me!' If that's the main purpose, then I don't think you should be posting that."

A selfie is never just a selfie, for Elise. It is an expression of who you are. You may look pretty, but if you are too eager to show that off, you might unwittingly reveal an ugly side of yourself.

THE SELFIE LOVERS AND THE SELFIE HATERS

Not everyone gave selfies as much thought as Tanuja and Elise, but everyone had an opinion about them—usually a strong one. When I asked about selfies, students tended to have one of two reactions. They would either roll their eyes and say, "I *hate* selfies!" or would light up and say, "I love selfies!" This held true in the online survey as well. A total of 364 students chose to give their opinion on selfies and our seemingly bottomless need to document our lives, and the question garnered some of the lengthiest answers across the entire survey. It was common to have one student start off his or her answer with "I *love* selfies!" or "Selfies are wonderful!" while the very next student began with "I *hate* selfies!" Students used more exclamation points in their answers here than anywhere else in the survey. Even those students with mixed feelings about selfies could swing from "Well the great thing about selfies is . . ." straight to "But then again, I loathe the way that selfies . . ." in the very next sentence.

One selfie-hater I interviewed, a young man, simply tells me, "They're stupid." When I ask why, he elaborates: "I don't need to see what you're wearing every day. I don't need to see a picture of your face every single day when I'm going to see you in ten minutes. I mean, if you're going out on a date or something and you're like, 'Oh, I look real nice today,' yeah, okay, it's a selfie. I've done that. I mean, if I look good, I'm dressed up for something, I'll take a picture of it and I'll post it on Instagram. That's different. But an everyday selfie? Come on. Like, come on! Let's be real. You have a little picture of your face up in the corner. If I want to see your face, I'll click on that. I don't need to see a different picture of your face every day."

This young man is not alone in feeling this way. Nearly twice as many students in the online survey were anti-selfie as were pro-selfie.

Many argued that selfies were "not good for people's self-esteem" or body image and that they contribute to the very problematic trend of "everyone constantly comparing themselves to everyone else" and driving people to compete against each other. Many students commented on how it was "very hard not to see it as narcissistic"; in fact, "narcissistic" was one of the most popular adjectives anti-selfie students used to describe the trend. A number of students chose this topic to express complaints about how, because of selfies, "we're no longer living in the moment," and this is a tragic loss. Selfies were called arrogant, self-absorbed, disgusting, degrading, ridiculous, vapid, useless, selfish, shameless, vain, and hedonistic. The general feeling seemed to be that the whole selfie trend has gotten "out of control" and people have "taken it too far."[1]

Several students were particularly harsh about what selfies say about their generation. A first-year student at an evangelical Christian college wrote, "I think what is driving our selfie culture is the millennial self-entitlement. We believe the world revolves around us and everyone cares about what we are doing. But the reality is, everyone is too busy caring about themselves to remember what someone posts for more than an hour or two."

Yet for many, the selfie is harmless and fun, a high-tech version of the glamour shots for which those of us who grew up in the 1980s had to pay good money in shopping malls.

One of the young men who lights up when I ask about selfies is a first-year at a Catholic university. "Selfies," he says with a sigh. "I'm a big selfie person. I love selfies!" When I ask why, he replies, "Everywhere I go I take selfies. Like I could be sitting in my dorm studying, and I get on my phone, and I have a Mac, so that's where I take my selfies. They're just funny. Especially if you're in class."

Attention professors and teachers: lots of students take selfies in class.

Another thing I learn from selfie-lovers is that there is a difference between selfies for Facebook and selfies for Snapchat. The difference is permanence, and it is huge. A "disappearing post" can be so much more fun than the ones that stick around and might haunt you for the rest of your life.

Jackson, a senior, loves selfies because "I get to see myself," he says simply, with a big laugh. He doesn't always share his selfies, though. "Sometimes I could be just looking at myself in the phone," he says. "Not really taking the selfie. And then I'll be like, 'Oh, I look nice today.'" Jackson likes using the selfie camera like a mirror, he tells me.

But when he does take a selfie, he has to decide where it goes: Snapchat? Instagram? Facebook? "I really don't care what I put on Snapchat," Jackson says. I could wake up one morning, upload that selfie, and I wouldn't put that on Instagram compared to my Snapchat, because [the morning selfie] is really more personal." The same goes for "sleek photos," which go up on Jackson's Instagram but not Snapchat. On Facebook or Instagram, Jackson also posts selfies of him doing positive things. Jackson works as a tutor, and this is something he wants to share with people in a more permanent way. "I might have a selfie doing some type of work with my students, some type of tutoring," he explains. If you got to Jackson's Facebook, you'll see that he's involved in community service. "But on my Snapchat," Jackson tells me, "you wouldn't even notice that I'm actually involved in community service because I'll be doing something funny."

The students I interviewed and surveyed constantly differentiated between permanent platforms and more temporary ones. Many self-proclaimed selfie-haters would later tell me how much they love Snapchat, using it to send seven or eight photos a day of themselves doing silly things. They didn't count these photos as selfies—maybe because the more "classic" understanding of the selfie, the one everyone likes to roll their eyes about, is that perfect vanity shot, the one that took a bunch of tries to get just right and is meant to go on your permanent Facebook or Instagram record. The Snapchat selfie is quite literally the "throwaway selfie" that isn't intended to last more than a few seconds. The Snapchat selfie is so fleeting it doesn't count.

Maybe Elise is right, and selfies really are all about intention.

THE SELFIE CONVERTS

Elise, like Tanuja, was once a selfie skeptic, but she had come around. There seemed to be a lot more of what I've come to think of as "selfie converts" out there among the other students I interviewed, too.

"I didn't like them at first, but now I've grown rather fond of them," one young woman told me. "At first it seemed very, very vain in a way. You would never catch me taking one. If I was in a public setting, on a train or in the middle of a crowd, I would never take a selfie. I would feel so judged. Now, it's fine. I don't know what changed. I guess I started doing it a lot more. I got *acclimated*. A lot more people who I thought were normal were doing them."

One of the most interesting selfie converts I met is a young man named Adam, who tells me his conversion story. He is excited to talk about social media because he recently starting dating his first girlfriend, and he is *in love* with her. He smiles the whole time he talks about her. Adam loves to share things about their relationship, especially selfies, and especially on Facebook, because he is proud to have a girlfriend. He wants to shout it from the rooftops. Having a girlfriend marks a new and exciting moment in Adam's life, and he wants her on his highlight reel. Adam is so sweet and enthusiastic that it warms my heart. Something he says is one of the most moving and positive things about social media that I hear in all my interviews, and it has to do with selfies. First, Adam makes a confession: he used to really hate looking at himself in the mirror—he didn't like what he saw.

"For the longest time, I'm like, 'Man! I don't look good,'" Adam admits. "But pretty much since my girlfriend, I actually look at myself like I'm happy with myself." This is where the conversation turns to selfies. "I never used to be a fan of selfies, and then once I started dating, I *love* the selfie," he says with a laugh. "It is not an official event with us until we have taken a selfie. [My girlfriend] made me so proud of the way I look and now I probably sound like a complete narcissist, but I don't know. I just enjoy it. I wanted to capture that moment."

Adam doesn't sound like a narcissist, though. His comment echoes that of Elise, who came to see selfies as a way to honor a moment when someone is feeling good about him or herself, creating a memento of it that a person can refer back to, perhaps on a darker-feeling day. And with Adam, his newfound interest and joy in taking selfies reflects and memorializes an important and healthy uptick in his self-esteem.

I ask Adam what he and his girlfriend do with all the selfies they take. "Upload 'em to Facebook and keep 'em," he answers with pride. They take all kinds of selfies. "Like, we did a charity event for cancer research.

We went to a nice dinner. We saw each other for something, I don't remember what. Then we went to the beach. Just, like, fancy trips, special dates." There's something about selfies, for Adam, that makes everything he does with his girlfriend seem more real. "I used to be opposed to [selfies], but once I actually started doing it, it is kind of nice because it puts a more solid base on a memory. Like, you can think something in your mind, but then it starts getting hazy. A photo, you can pull up and be, like, 'Oh, I remember this. That was happy, that's when we went to [a restaurant]. That's when we went to the beach. That's when we went to that.'"

I ask Adam whether the selfies he posts get a lot of "likes." "Yeah," he says. "A lot for me is, like, ten to fifteen 'likes.' It's my friends, and my family and her family. Her family really likes me."

In the online survey, about a fifth of the students had positive feelings about selfies, and the reason coincided with the opinions of Elise and Adam: that selfies can be good for boosting self-esteem and forging an identity (and particularly so for women).[2]

"I love the selfie culture," said one such student, a first-year woman at a private-secular university. "It's a movement to take your identity into your own hands. A lot of women are able to control the way they portray their own bodies and beauty, and I think that's wonderful." Other students commented on how selfies "reflect a positive body image" and "promote self-confidence and body positivity." Another young woman mentioned the political power of selfies: "Especially for young women and marginalized groups, selfie culture is super powerful." Yet, she wrote, because selfies are so often associated with young women, they also get unfairly critiqued.

Another woman who started off her answer with "I love selfie culture!" went on to list how good selfies are for self-acceptance. "It's important to have a place to express yourself and love yourself in a positive manner," she wrote. "Some people feel annoyed by selfie culture, but why? It's so rare to find self-acceptance. The goal is to love myself so much, others wonder why. Happiness is found in self-love. It's important to celebrate yourself!"

Yet another pro-selfie answer came from a woman who saw selfies as a way to affirm and "honor ourselves," even if others don't think we are attractive. "It is a way for us to display pride in ourselves, even if we are not good looking, because we did the act of taking it," she wrote. "Before, others would honor us by taking our picture, and they still do. But given technology, we can honor ourselves without the community of another individual desiring to take our picture."

Selfies seem to offer these students an important opportunity for self-affirmation, especially among those who feel this sort of affirmation from others is typically lacking in their experience. The act of taking a selfie—deciding for yourself that your image is worth capturing and posting for others to see—is a way for some young adults to take control back from their peers. Rather than letting others decide who and what is photo-worthy, selfies allow a person to decide this for him or herself.

One young woman even found selfies inspiring, and helped to remind her of her future dreams. "For some people, like myself for example, seeing pictures of people traveling and living the American dream, it pushes me to want to finish school so I will be able to provide that life for my future family," she wrote. Another reflected on what aspects of the self are fulfilled by selfies: "We like seeing our own face. There's a little bit of vanity behind it all. Also playfulness. And adventure."

Even students who take lots of selfies and love them know there is a limit on posting them, though—if you post too many, people will think negatively of you, so you have to be careful how often you do it. But most people will forgive a selfie now and then, because they've probably taken one at some point, too.

SELFIE CULTURE

When I asked students what selfies are all about—What do they think started the trend of documenting ourselves this way? Where is that impulse coming from?—many of them weren't sure what to say. They were sort of miffed about the trend and the origin of the impulse, even though so many of them participate all the time.[3] Some people simply cited the ease of taking pictures on smartphones and the invention of the front-facing camera.

However, one young woman, Amy, gives me a kind of chicken-and-egg response. "I think it's kind of a snowball effect," Amy says. "Because I don't think this used to happen. I mean, I'm younger and this has always been a part of my life, but I tried to picture, you know, what it was like before the Facebook/Instagram thing, and I don't think everyone was showing how awesome they are in your face all the time, or at least I hope not. But I think, like I said, things kind of snowballed because people see other people sharing their life, and then they feel the need to contribute to that, to make it seem like their lives are really cool, and then they do things to, you know, keep up with the Joneses. I think other people see that, and then they join in, and I think it kind of just spirals out of control and becomes a thing where everyone's doing it, and everyone's constantly feeding into it."

There are also students who genuinely worry about selfies and the way people are becoming obsessed with posting photos about every little thing they're doing as a way to "prove" they're doing it—"pics or it didn't happen," as the popular phrase goes. Students worry that selfies are taking priority over living one's life—that people do things for the photo and not for the experience. And some students, like Max, who attends an evangelical Christian university in the Southwest, think that selfies and this impulse to document play into the worries about future employers and the need to provide them with a highlight reel.

"I think a lot of people try to prove to themselves and to others on social media that they are someone that they're not, or that they're trying to show that they are having a good life," Max says. "This veil of not being authentic, I think it comes from the very root of why you're taking these pictures, to put [it] in very literal terms. For me, my personal account, I enjoy photography so I just enjoy sharing that. I think that when I share pictures, it's not to make other people jealous. It's not to make other people, ooh and aah at what I'm posting. It's not to show that I'm hanging out with certain people when it comes to little things, like who you tag in a picture. I think that has a lot to do with self-worth, in that kind of culture." Max thinks that selfies can make us feel really insecure because they offer idealized versions of people's lives.

Max thinks that the "photo culture" we see online has a lot to do with professional concerns, and even here, in a conversation about selfies, Max adopts the language of business as he tries to explain the phenomenon: "Social media is a powerful tool for promoting the self. If we're going to go in terms of companies, it's a promotion tool, it's a marketing tool, it's those kinds of things. So there's that facet of it. In terms of a personal account, whether it be Facebook or Snapchat or Instagram, I think it's, like, someone's ideal self sometimes. They put too much weight into the way they view themselves, the whole selfie thing."

Selfies are a big part of the happiness effect. They are a product of the need, even the perceived requirement among college students, for self-promotion online. And this in turn takes away from the real joys and happiness of living something without worrying about having to promote it on social media.

"I personally have a lot of beef with it," Max says. "I think that you lose the intimacy and the importance of the moment when, the second you pull out your phone to take a picture of it. Even for me, if I think something is awesome, whether it be a hike I just did or a concert I just saw, those kinds of things, I make sure to enjoy it first. The reason that I'm going is because I enjoy those things, and then I take a picture more for the memory of it. You know, I think that I can look back on my Instagram account twenty years from now, if that app still exists, and be fond about these memories, more so than [to see] what I was trying to promote or what I was trying to make others see, or make myself seem like I was happy, or I was fulfilled." Max sees his photos as a tool for remembering experiences, but he thinks that for other people selfies are replacing experiences. Many other students made similar comments, sharing Max's worry.

"I think that at the root of documenting experiences is insecurity. We feel this inherent need to post things that will validate our lives," Max tells me. "It's become this competition of who can go to the coolest places, who can find the coolest things, who can take the best picture of themselves, and it just becomes this very selfish, self-centered culture that I think is taking our society in a wrong direction. Because people are starting to lose out on the importance of the events themselves, whether it be a concert or a hike, those things that I was referencing. People aren't enjoying the music anymore, they're making sure to get the perfect angle. I find myself in those positions as well, but I think people take it to an unhealthy extent."

As is apparent from his comments, the culture of selfies is complicated for Max: he takes selfies, but he also worries about them.

David, by contrast, is not at all conflicted. David does not "do" selfies. "I'm *not* gonna post a selfie everywhere I am," he says forcefully. "I wouldn't post a selfie on my way walking in. Like, 'Oh, I'm about to go to an interview,'" he jokes. Then David effects a funny, falsetto voice. "I want to post a selfie right now!" He seems disgusted with the very idea that people take pictures of themselves all the time.

David searches for a reason behind the trend and comes up with a number of possibilities. "A lot of people, you know, just do it to update people on what they're doing, and I also blame our technology a lot on it, our new phones," David explains. "You have a great camera. Why not take a new picture every day on your new iPhone 6? Might as well!"

David is passionate about this. "I'm *so* against it," he emphasizes. "Personally, if I go on Twitter or Instagram and I see the same person every day, it makes me mad. Why do you post stuff every day? It's cool to know what you're doing, but I don't want to see you on my timeline every day." David likes that he can keep up with people on social media, but it *really* bothers him how constant the influx of information is. "Don't put enough for me to know what you're doing, where you're at, you know, who you're with every day," he says. If David wants more information about someone, he feels he should just go directly to the source and ask for it.

According to David, women do this more than men—though men do it, too. "Women are posting *a lot* of selfies," he says with a laugh. "And guys don't post as much, but guys still do post selfies." David thinks that girls simply post more than guys. "Guess they're more interested in social media and taking pictures," David comments by way of explanation.

SOCIAL MEDIA: A WOMAN'S WORLD?

David has a lot of company in thinking that women post more selfies than men, and that women post more often on social media in general. The majority of interviewees seemed to think that women take and post more selfies than men do—by a lot—and most think that social media is more of a "girl thing." Both women and men feel this way.[4]

Cherese, she of the seventeen Facebook groups, is sure this is true. "First off, guys really don't post too many selfies," she says. "They won't post too many selfies, and all theirs will be just all different, like, a football game or it'd be something maybe sports related, but it would be nothing that's actually personal. As opposed to, like a gay man, or women, who post more personal things about themselves. They may post about their whereabouts or how they're feeling, more of the emotional side, as opposed to just generic things that are happening in the world that anybody could've seen."

In this short statement, Cherese hits on most of the major gender differences students perceive:

1. Men don't post as many selfies.
2. Men post about sports, cars, and events happening in the world.
3. Men don't post about personal things in their lives.
4. Women almost always post about personal matters and their emotions.
5. Women post selfies and lots of pictures in general.

Cherese's comment about how "gay men" also post in the same way as women was not something I heard often, though it did come up occasionally. But what Cherese is saying about women is an expression of age-old gender stereotypes: women care about appearance, emotions, and the personal, whereas men are about being active and out and about in the world. These are the very same stereotypes that scholars Lyn Mikel Brown and Carol Gilligan documented and critiqued in their groundbreaking book from the early 1990s, *Meeting at the Crossroads: Women's Psychology and Girls' Development*, which looks at how, when girls reach adolescence, their worlds grow smaller, less active, and become about image and appearance, whereas boys' lives become about their important place and future in the public sphere, as well as all the amazing things they will do in the world (to sum it up succinctly: boys' lives expand, and girls' lives contract).[5] Decades later, these very same gender norms are playing out on the social media profiles of college students, with both women and men rattling off these norms with very little (if any) critical concern about the fact that with this commentary they are playing right into these stereotypes. These students seem to take it for granted that women are more into selfies because selfies show off their faces and bodies, and men are not because they don't have to care as much about that sort of thing and instead can focus on their exciting adventures and the occasional newstory—as though this is simply the way things are and how the world works. Whether or not students can identify these as gender stereotypes, they certainly see these classic stereotypes playing out on social media every day and report back what they're seeing accordingly.[6]

Matthew, who goes to the same university as Cherese, echoes her comments and then adds a few to the list. "Guys usually post more athletic stuff," he says. "If they want to post something funny, then it's usually more vulgar, I suppose. And if a girl is posting something they think is funny, it usually has to do with, I don't know, something weird that their friend just did. Like more innocent, I guess. As far as the food thing goes, it seems like it's almost always girls posting what they're eating." Matthew pauses to laugh here. "This probably sounds bad, but I feel like, on Twitter at least, there's definitely more complaining from the girls that I follow about things than there are from the guys. . . . My friends play this game where we'll read off somebody's tweet and we'll try to guess whose girlfriend posted that or which girl from the high school said that because they're always real similar."

Matthew hits on another of the stereotypes about "what girls post" that I hear frequently: girls refer to their friends and are more relational, more expressive, and more willing to share. Matthew and his friends see girls' expressiveness as whining, in some cases.

Elise really sums up the notion that social media is a women's world. "I don't see guys post that much, actually," she says. "I think it's definitely more of a girl thing. I don't know exactly why that is. Maybe girls are just more open with their feelings and, you know, their thoughts of the day, I'm not sure. But guys are definitely more just, more *blunt*, I guess. . . . Whereas girls, you know they'd put all the emojis and all these different hashtags and things. So I think girls are more open on social media because I think because the majority of people that post are girls. I think it makes it easier for girls to connect with other girls they don't know, or even girls that they hardly know. And guys seem, especially on Instagram, like they only comment on pictures if it's a girl that looks really pretty. And, I mean, that's just a guy thing, but you don't see guys putting up selfies that much."

Most students also seem to accept the stereotype that women are "more relational" than men (whereas men are all about themselves), and because social media is supposed to be a sphere for connecting to others, students think it's simply more of a woman's world. One piece of anecdotal evidence supports this notion: the online survey—unlike the interview process, which involved random sampling—was all voluntary. Students self-selected to take it, presumably because they had something to say about social media or because the topic interested them. And three-quarters of those who completed the survey were women, which means that, when given the chance, women were more likely to share their opinion about social media. Is this simply because (to return to stereotypes) women are more expressive and so are more likely to complete a survey? Or is it perhaps because women are actually more involved and invested in social media? According to a survey by the Pew Research Center, women absolutely are. And the statistics show that "girls dominate social media," whereas "boys are more likely to play video games."[7]

These stereotypes are clear in the answers students gave to the online survey question asking if they noticed gender stereotypes in the ways people post on social media. Eighty-five percent of the students who replied to this question noticed at least some differences in gender, as indicated in Figure 4.1.

The essay responses to this survey question echo what I heard repeatedly during the interviews about gender and social media. Women are more expressive overall, in the students' opinions. Women are more relationship-centric, and their expressions are all about appearance, showing images of themselves, displaying their bodies, and getting other people to notice their looks. Women are more about selfies and are more

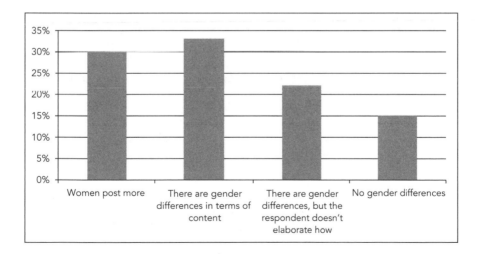

Figure 4.1 | Student Replies about Gender Differences

self-centered in general, as well as more likely than men to seek attention (often pathetically so, according to some students). Among those who believe that women and men post the same amount—just about different things—students often explicitly mentioned they believe that typical gender stereotypes apply on social media. Many of them spoke of how unfair gender stereotypes are to women, and how it's unfortunate that women have to deal with these same stereotypes on social media as well as elsewhere. These students believe women are judged unfairly in general because of such stereotypes and face those same unfair judgments playing out on social media as well.

For this essay question, students' answers reflected the same kinds of ideas about men that I heard during the interviews. Respondents wrote of how guys are all about showing their life interests, displaying their hobbies or themselves playing sports, and putting up photos not so much of themselves or their friends but of things they like—travel, cars, and sports being the most popular. For men, social media is about displaying your successes and the things you do in life. Men are considered bold and outspoken, more honest, and funny. Guys get to be more "careless" or "carefree" on social media because overall in our society guys are judged more fairly than women; if men put up an offending post, they probably can get away with it without major negative repercussions, whereas women will get punished for any and all mistakes. A number of students felt that guys were clearly more interested in sex, dating, and hitting on women, and occasionally being vulgar about women on social media— something that didn't come up at all with regard to what women post.

Over and over, students' comments on this subject seem to confirm (at the very least) their belief that social media is more of a woman's world. And when students are discussing why they believe this is the case, the conversation invevitably cycles back to talk of selfies.

Laura, for example, is a selfie-hater who thinks selfies are "so dumb." What's worse, though, is that she thinks women are the ones being "so dumb" by taking them, especially "naked pictures" or just "very sexual pictures," in an attempt to get attention and approval on social media. Girls take sexual selfies because they want "likes" and want people to affirm for them, again and again, that they're beautiful, and they do so in public in front of everyone they know. "I just don't like [selfies] and it aggravates me when there's thousands of them on Facebook," Laura says.

Laura, like many women I interview, judges other women's activities on social media rather severely, yet is far kinder and more forgiving when talking about men and how they operate online—at least at first.

In Laura's opinion, you just don't see guys putting up selfies. She cites the example of a friend of hers who is extremely handsome. "He's a great-looking dude, but you would never know that off of his Facebook, because he just doesn't post anything," she tells me. "He doesn't care, but girls make sure they have the best picture because they think guys are going on there, which they may, I don't really know." Laura thinks women simply care more about social media, but then she goes on to express a bit of a double standard about the men now, too. "Guys don't really care that much about Facebook and making sure that they have a good Facebook, and then *when they do*, you're like, 'Why do you care so much about yourself? Why are you so arrogant? He's taking really good pictures of himself. Why does he think so highly of himself?'"

Although Laura thinks it's ridiculous for women to post selfies, she perceives it as "normal girl behavior"—it's just more of a girl thing to do. But when a guy *stoops* to such behavior, it's even worse. Men aren't known to participate on social media in the way women do, so when you see one engaging in what is stereotyped as "typical online feminine behavior," he gets judged extremely harshly—as harshly as Laura (and her peers) will judge another woman. The worst thing a guy can do on social media is act like women do when they are on it.

Ian, a junior at a private university in the Midwest, acknowledges straightaway that he will be defaulting to stereotypes in order to discuss gender on social media. "With girls, obviously these are blatant stereotypes, there seems to be kind of two categories. Avid Facebook user—*that's* a girl," he says, by way of starting out. "There's the one who touches up all of her photos using Instagram or, you know, Photoshop or whatever, maybe adjust the lighting a little bit, so everybody looks really nice in the photos or something, and posts tons and tons, you know, forty photos from a night. You know, four of them are the same group photo, someone's looking like this and then like that, and then, maybe some other person got pulled into the photo, but it's like, *man*, one group photo, one big group photo would have been enough to say who was there, that it was a party and all the information. And you have to scroll through thirty permutations of the same photo really."

That's just one type of girl Ian sees on Facebook. Ian goes on to describe the other. "And then you get the girls who are very self-deprecating and will post a silly photo of themselves from their childhood," he says. "My girlfriend's good friend is big on this. She's lost weight recently, but she used to post these photos of herself from sixth, seventh grade, you know, when she had big red cheeks and was real chubby and all of that. It gets tons and tons of love and 'likes,' and everybody's laughing and having a good time, but it's almost like a *produced* candidness. It's like, you're seeing a goofy look of me, but you're not going to think any less of me. It's because I realize this is silly of me as well."

For Ian, like others, social media is very much a "girl thing" because women are appearance-conscious, and women like posting photos of themselves looking good and partying with their friends. Either this, or women are self-deprecating and silly, which is also, in his view, about appearances. For Ian, women are not to be taken seriously in general, and certainly not on social media.

Ian was blunt about employing gender stereotypes, but some other students were incredibly nuanced and eloquent on the topic. One such young woman, who was very pro-selfie, went on at length about how, in her opinion, negative judgments about selfies have everything to do with negative judgments about women and stereotypes about gender and race. "I've noticed that when we make fun of selfie culture, our jokes are intertwined with jokes about white girls and valley girls, which is something that I'm pretty bitter about," she wrote. "I get the feeling that if more guys than girls took selfies, the jokes wouldn't be so popular, and we wouldn't look down on the selfie takers as much. This might be linked to the fact that girls are encouraged to care about their looks by every aspect of popular culture, but are also encouraged to be humble and somewhat modest by the exact same entities."

Not everyone feels this way, though. In fact, some people believe that social media—and in particular the trend of taking selfies—is making guys care about their appearances and their bodies in ways they haven't before. Tara, for instance, tells me that her opinion on gender, selfies, and social media has changed recently.

"If you asked me this a couple years ago, I'd say, 'Yeah, only girls post selfies,' but there are a lot of males that act the exact same way on Instagram and Twitter," Tara says. "Instagram was a girlie thing, but now it's really not. A lot of guys are on it, and they essentially post the same sunset, same workout pictures, that kind of thing." Tara thinks the difference is simply that Instagram's popularity has shot up and that the apps are easy to download on people's phones. "I think it's more, just, the image of Instagram developed and became bigger. Now basically everyone's like, 'Yeah, I'll get it' and download [it] on their phone."

But Joe, a sophomore at a private-secular university, happens to be a power lifter, and this changes his entire experience of selfies and social media, though he starts off with a joke about guys and selfies.

"I'm going to preface [my answer] with a really quick comic that I saw, it's like, 'How guys should take selfies,'" he says. "It's like, 'First, pick up your phone. Then lower it. Lower it even more. Set it on the table. Guys don't take selfies.'" Joe laughs at the joke before continuing, telling me, "So I've never been a super big person taking selfies and stuff." Then Joe turns to the subject of selfies and women. "I think the problem with selfies is that they can promote vanity to some extent. You spend twenty-five minutes getting your makeup ready, just so you can take a snap to send to your boyfriend or a guy you're crushing on. Then beyond the vanity aspect, they do promote more of a superficial image of who you are." But, for Joe, selfies don't always have to be superficial. Sometimes a selfie is just a nice picture. The problem, he says, is "when you see somebody who's purposely posing in a certain manner, you know, positioning themself in a certain way, facial expression, et cetera, I think it portrays a certain mentality. An example of that would be the duck face. You know? With certain females, I've even seen some guys [do it]."

I heard a lot about the infamous "duck face" selfies that women take—lips puckered, cheeks sucked in. This pose is particular to women, according to just about everyone, though Joe has "even seem some guys" post such photos. People on campuses with Greek life seem to think that making the "duck face" was typical sorority behavior. It is a style of selfie that students like to make fun of, but that no one admits to taking themselves.

Joe isn't done talking about "duck face" selfies. He continues, "If you see a girl posing with the 'duck face,' you're like, 'Oh this girl is very vapid. She's not very intelligent. She's superficial. She's very into how she looks and her clothes.' And she may be an amazing, great individual, very bubbly, very talkative person. You're not going to get that from the selfie that you just saw. You're just going to see her as being whatever society has attached to the 'duck face.' And what society has attached to the 'duck face' is, it's a primarily suburban white girl type of thing, who's not exceptionally intelligent, she likes to drink you know, lattes, wear Ugg boots, and she's kind of stupid. And so I think when you see that, unintentionally, even if you're aware of it, you may have some stereotypes preconceived."

After Joe finishes this speech, he shifts direction on the subject of gender and social media. "To be completely honest," he says, "I have not noticed too much of a difference" between men and women. Women put up photos, but guys put up photos, too, and people of both genders react. "It's not like girls are the only ones complimenting. Girls compliment pictures. Guys compliment pictures. Girls will say, 'I just went to the store.' Guys will say, 'I just went to the store.' Girls will say, 'Look, I'm working on this project, you should check it out guys.' And guys will also say the same thing."

Many students raise the subject of how a *woman's* body and self-image are affected by all those perfect selfies (and the bikini photo, if someone dares post one); very few students talk about how men are just as affected by having all these images coming at everyone all the time—but Joe is one of them.

"Before I got into power lifting I was into bodybuilding, and I followed a lot of bodybuilders on Facebook and stuff," Joe tells me. "All these different guys who are either body builders or physique models. And they'd always have pictures. And I would see these pictures and it's like, 'Oh wow! I'm really aiming to have a really big back like this guy. I really want a nice peak on my bicep like this guy. Look at that guy's calves. He's really cut up. He's like 4 percent body fat. I wonder if I can get to that.'" For Joe, the photos are inspiring. Rather than making him feel inferior, they make him want to meet his bodybuilding goals. But he credits his ability to not let these photos get him down to "being a relatively well-informed individual and to some degree intelligent," he says. "I think that I can objectively look at these postings by these people and say, you know what, I'm going to work toward that goal. I will most likely never get to where they are at. I'm going to become the best version of me that I can be. And I'm going to do everything I can. Maybe one day I will be close to where they're at, but I don't think that I could ever get there. Now that's coming from the perspective of somebody who is well-informed and is aware."

But not everyone is as self-aware as Joe thinks he is, or as comfortable about their own body and its limits. "People who may be less aware"—Joe hesitates here before moving forward—"I don't mean to stereotype anybody, the average American, the average American high schooler, they are heavily influenced on body image by seeing these things. If you had somebody seeing all these postings by athletes with great physiques, they would be like, 'Wow,' you know, 'I need to have a great physique' or 'That's what I want to work toward because that guy's really cool.' Other people are saying, 'He must be cool. I want to be cool. I've got to look this way. I've got to make myself into what this guy is.'"

If resisting the urge to wish you were different is hard for "the average American guy," Joe thinks, the images people are exposed to on social media are far harder on women. "I think even more than with guys, with girls there's an incredible, incredible, *incredible* social pressure to fit that Barbie doll model," he says. "And for guys, too, it's completely ridiculous to see all these pictures of these Photoshopped, touched-up women . . . . and it creates this completely unrealistic expectation for both sexes on what their beauty is supposed to be. So girls will spend two hours, three hours in the morning putting their makeup on to go to school because they can't have a zit, a single zit on their forehead showing, or a little red spot on their cheek because if they do, that'll show to others that they're not perfect and that they're therefore not as desirable as somebody else . . . . and so I think that body image is very warped by social media especially in the eye of the uninformed or the less informed individual."

While social media may indeed be more of a "girl's world" in the sense that women participate more regularly, it certainly doesn't seem to be a fairer one as far as women (and women's bodies and appearance) are concerned.[8] This makes social media yet another sphere where women experience sexism in our culture, and where men, ultimately, have it easier than women do. Young women not only must live up to expectations around professionalism and image-building but also must look good doing it. They have to walk an impossible line between being sweet and innocent yet also sexy (at least to a degree). Missteps can cost them both socially and in their future careers. Overall, college students feel pressured to maintain a presence on social media, yet because social media is really "for girls" and those girls can be oh-so-annoying when they post too much, women have to be especially careful about how they post because they are being scrutinized more closely by everyone else (including—maybe even especially—by other women).

The trend toward taking selfies and documenting everything people do is not just about gender stereotypes, but there's no doubt it exacerbates them. Women—young women especially—are expected to offer up their images not only for viewing but also for evaluation. Girls and women have always been judged by their appearances, but social media takes this tendency to a new level of intensity and constancy. And while stereotypes about gender and social media abound among college students, the institutions they attend do not seem to be taking this issue seriously enough to unpack and critically analyze such biases in the classroom. So they simply rage on.

THE FUTURE OF THE SELFIE

So, fifty years from now, will people be calling this the "Selfie Generation," as Tanuja wondered? The odds seem pretty good. Tanuja might want to trademark the phrase. And while the "Selfie Generation" may be new, the gender stereotypes are very, very old.

But if that phrase were to become shorthand, implying that today's young people are all entitled narcissists, that would be very far from the truth. Some selfies may be expressions of narcissism, but many are merely playful, self-affirming (and even empowering) creative expressions of identity, in a sphere where playfulness and creativity attached to one's real name is at a premium. The downside of this trend, however, is worrisome. It seems to foster overwhelming anxiety among young people of all varieties, especially young women. It feeds right into college students' anxieties about documenting certain highlights about their lives for a particular audience, as well as the importance of curating a particular image online. Just as the professionalization of social media has exacerbated the dissonance between the real and online selves, the trend toward taking selfies and the pressure to constantly prove what one is doing and with whom (remember: "pics or it didn't happen!") interrupt the real lives and experiences of students with ever greater frequency. Young people feel anxiety and stress around what they perceive as fakery in how they and others present themselves online. Yet with the pressure to post photos to *prove* that their lives are great comes a further intrusion into time spent alone, with friends, and participating in the activities they enjoy. Whatever else selfies are, they are a particularly immediate and egregious cause of the happiness effect.

While so much of social media has become about performing for an audience—whether that audience is comprised of your peers, your coaches, your parents, your sorority sisters, or your future employers, for some young adults, there is a much more important audience than all of the above: God. Religion is one of the most central and intimate parts of a person's identity. If social media is the main way that young people today express that identity, it naturally raises the question: Where is God on Facebook?

5

PERFORMING FOR GOD

Religion On (and Off) Social Media

Facebook keeps me in check with the secular world. Which is very important because I go to a secular university, I live in the secular world.

Zachary, sophomore,
private-secular university

The Internet makes everything else easier to do. It's easier to find recipes, it's easier to find songs, it's easier to find other people who have the same crazy chronic medical conditions as you do. It's easy to find everything so of course it's easier to find boys. It's a side effect of the Internet which a lot of people don't like, and that's why there are some people who are Orthodox who don't use Facebook.

Dinah, senior,
public university

JENNIFER: GOD USES FACEBOOK

From the moment Jennifer sits down for our interview, I know I'm in for a treat. She's a bright, bubbly senior at a conservative, southern, Christian university. A pretty redhead with freckles, she talks enthusiastically about all the things she loves about her studies, her experience at college (she's made two "lifelong friends," she immediately tells me), and how, during her four years here, she's been "pushed in the best of ways." She has a ring on her finger, too—she's engaged and thrilled about it. At the core of all these things, for Jennifer, is her faith. She's a devout Christian and a member of the Pentecostal Church of God, and she can't really talk about anything without bringing up her faith. Her father is a pastor, and she both teaches Sunday school to preschoolers and works with her church's youth group on Wednesdays. Jennifer is a psychology major, and while this has challenged her faith at times, she believes that her major has helped to "expand her horizons."

"[Religion] is a big part of my life because I do have a relationship with God," Jennifer says. "You know, we talk on a daily basis, and it definitely influences the decisions that I make. My faith is a big part of how I decide things I'm going to do, things I'm not going to do. I don't think I worry as much as other people do because I believe that there's a plan and there's a purpose for everything that's destined by God. So I don't have to worry about the future as much, because even though it's unforeseen, I know that everything will be okay."

When I ask Jennifer what makes her happy and what makes her life seem meaningful, our conversation again turns to faith. "I know that I have a purpose, based on my faith in God," she says. "I know that He created me to have a purpose for this life, and as long as I trust and obey Him, then He'll make sure that His plan for my life is seen, or He'll guide me into that plan He has for me."

I begin to grow curious about the ways Jennifer's Christianity will affect who she is online and what she posts, if at all. Most of the students I've interviewed are only nominally religious, so faith isn't something they discuss when they're talking about social media. And the more devout students I've met usually tell me that social media isn't the place to talk about religion, and just generally students see posts that have anything to do with God, prayer, or even worship attendance as one of the biggest no-nos of online behavior, right alongside politics. It's the old cliché about polite conversation updated for the twenty-first century. It's best not to ruffle any feathers on social media—you never know who might be watching.

Even Cherese, who enjoyed expressing her opinions about everything—even politics and religion—on Facebook, spent nearly all her time navigating the seventeen groups she'd created to protect almost everyone she was friends with from seeing these posts.

But Jennifer is *very* different. At first, she sounds like most everyone else, saying that she never really posts anything bad on social media— if you have a bad day, you should keep it to yourself. She posts happy things and things she's thankful for. Jennifer says that one of the things she asks herself before she posts is whether or not what she writes will "uplift" others. She wants to make other people happy; she believes people will be happy for her when she's doing well, and she also gets pleasure out of seeing others doing well. Jennifer gushes about the announcement she and her fiancé made about their engagement and how many "likes" and sweet comments they got. It was her most popular, most uplifting post ever—exactly the kind of thing that belongs on social media, in her opinion.

Then I ask Jennifer if she's open about her faith on social media. Yes, she tells me, she definitely is. At first the posts Jennifer mentions that relate to her faith seem pretty low-key. If she finds twenty dollars, she'll post something like, "Lord bless me today, I found twenty dollars." Sometimes she'll ask for prayers on Facebook, too, usually for things like an exam she's studying for or a paper she needs to get done, but never for personal or emotional needs. "I don't think the whole world needs to see [those things]," she says. "There's a lot of people on Facebook who, I don't want them to know my needs. I don't confide in them like that. The people who I would request prayers from for personal needs or emotional situations, I would confront face to face, usually one-on-one."

This is where our conversation circles back to Jennifer's earlier mention of God's plan. "I think God can use anything for His glory," she says. "Not everything on Earth may have been *set up* or *created* to glorify Him, but I think He can use it. So, I think you can spread the gospel, you can spread hope, you can spread love through social media, which would be, by doing that, you would be glorifying God and you would be uplifting Him. So I think He can use social media."

I ask Jennifer to explain what exactly she means.

"Yeah, yeah," she says, laughing. "I mean, not literally, you know. He doesn't sit down and type up a message for you, or whatever. But, yeah, I think through people, because people are the ones that are using Facebook, and God can use people, so if He can use people to do anything, to preach in a pulpit, to sing a song, He can use them to type an encouraging message on Facebook, or spread hope or love through a post that could uplift others, or encourage them to continue, you know, the fight or whatever." It's not as if social media is holy, Jennifer wants to reassure me, it's just that the Lord works in mysterious ways and sometimes it's through Mark Zuckerberg, even if he doesn't realize it. "I'm pretty sure [social media] wasn't created to glorify God, but it doesn't really matter, I don't think, because God can use anything."

Jennifer goes on to say that social media can also be a distraction from God if you're not careful, and anything that distracts from God leads to unhappiness. "I think a lot of things can pull people away from God," she begins. "I think social media can play a larger role in pulling people away from God because it's such a distraction with a lot of things that aren't Godly. There's a lot of things on social media that are not there to glorify God, to uplift Him, or to encourage people. There's a lot of things there that can tear others down. People post things that the whole wide world doesn't need to know. People put up situations that are discouraging. People post videos that can be hurtful. But there's also just a lot of other sinful things too. Like language and videos, crude jokes, crude humor, that's not glorifying God. So I think *that* can pull people away." And while seeing hurtful, crude, or discouraging things on social media isn't much different from seeing or hearing about them in person, "social media is at their fingertips all the time," Jennifer says. "But once again, it's not necessarily the media that's doing it, it's the people behind it. It's the people that are posting those videos, posting those posts."

While so many of Jennifer's peers are working hard to please and impress their audience—future employers, professors, college administration, grandparents, sorority sisters, and fraternity brothers—Jennifer's number one viewer is God. She believes God is always watching her, even on social media, so she posts accordingly, trying to appear happy at all times and to inspire others, but most of all to serve God.

When Jennifer goes online, she tries to ask herself, "Will this glorify God when I post this?" She doesn't *always* do it, she tells me. She's human, so she makes mistakes. But she believes that when she's on Facebook she's in a "role model state," so she feels a responsibility to do the best she possibly can to allow God to work through her.

The difference between Jennifer and almost everyone else, though, is that the effort to please God and appear happy doesn't seem to exhaust her. Rather, she seems invigorated by it.

JAE: BEWARE OF FALSE IDOLS

At this same Christian university, I meet Jae. A tall, lanky junior, Jae is quick to laugh and spends a lot of his time on Instagram and Snapchat—like everyone else at his college, he says—but he also likes to go on a Korean social media platform. Jae is in ROTC, which is where he met a lot of his friends, but he has a separate friend group made up of people who identify as "part Asian," he explains. Jae is very religious. He prays regularly, and it's important to him that his future wife is a Christian. He tells me that God makes him happy, that God is his number one priority, so serving God is what brings meaning to his life.

As with Jennifer, Jae's relationship with God and his devotion to his faith have an enormous effect on his use of social media. He goes on Instagram and his Korean app quite a bit, though he doesn't post much. When he does, he's either trying to say something funny or putting up something that reminds him of God. "If I post a beautiful sunset or something and then I talk about how it reminds me of God's beauty, it would help not only express my own gratitude of God, but it can help show other people who look at my page, like, 'Yeah, he's a Christian,' and [it's] a good way to get the word out." Jae likes being able to remind people of the wonders of God through his posts.

But while Jennifer believes that God can use anything to glorify God's self, and hopes that God works through her on social media, Jae worries that social media has the potential to get in the way of his relationship with God or even destroy it, if he's not careful. In fact, he once quit Facebook precisely because he worried he was becoming obsessed with it, and obsessions are harmful to a person's relationship with God. Social media can consume you and make you forget your priorities.

"Social media can be like another God to people," Jae says. "It can be idolizing, and that's not what I want, so it affects how much I go on."

When I ask Jae why he thinks social media can be "idolizing"—in the religious sense—he talks about how tempting social media is to people. He explains, "People can be anything they want, they can do whatever, they can make multiple personalities of themselves that they can't do or won't do in their real life." For just these reasons, Jae found himself fixated on Facebook for a while. He used to spend a lot of time scrolling through everyone's feeds and obsessing over how many "likes" he'd get on his posts. Then, Jae started to feel emotionally unhealthy because of Facebook. He kept seeing posts he didn't want to see: negative comments, negative images, and people talking badly about other people. Jae got to a point where he just "didn't want to have those images or those thoughts in his mind throughout the day." Plus, "I didn't want to fall into the idolizing trap," he says, where Facebook becomes like a God to him.

So one day, he quit, cold turkey. He tells me, "I was just like, 'All right God, I see this is not going to be beneficial to me, for our relationship,' so I just cut it off."

Sometimes Jae misses Facebook, though. He worries he doesn't really know what's going on with a lot of his friends, or what's happening on campus. He follows some people on Instagram, but Instagram isn't the same, he thinks. Lots of people aren't on Instagram. When I ask Jae whether he ever worries that Instagram will consume him the way Facebook did—becoming yet another false idol—he answers with a simple nod of his head and says, "Yes."

SOCIAL MEDIA EVANGELIZATION

College students do not generally reveal their religious preferences, affiliations, feelings, or opinions on social media. At the thirteen colleges and universities that I visited—even the religiously affiliated ones—people like Jae and Jennifer were anomalies. Very few students posted about religion or spirituality.[1] A small number admitted they occasionally posted a Bible verse, along with a photo of nature, or made a comment about praying for someone who is sick—certainly more so at the evangelical Christian colleges than anywhere else—but that's about as far as their revelations go. Nobody wants to start a debate or cause a ruckus, and posting about religion is inflammatory, most students think.

In the online survey, students were asked to reply yes or no to the following statement:

I share my opinions openly about subjects like politics and religion.

Only 25 percent of those who answered said yes. But those 25 percent are fascinating.[2]

What is unusual about them is that they are, more so than their peers, empowered in their use of social media. Those students who allow the devotion to their faith to permeate their online worlds use their religious traditions as a framework for navigating their behavior and posts— one they find far more meaningful and sturdier than warnings about future employers and prescriptions for curating one's online image. They are learning the dos and don'ts of social media from a higher power, and this makes an enormous difference. Just as students at the most prestigious institution I visited, who thought critically about social media, had a healthier relationship to it, navigating social media via one's religious tradition also seems to give students who do so a heightened sense of control and purpose. And while these students are just as image-conscious and as aware as everyone else that they have an audience, having God and their faith tradition filtering their online decision-making seems to help them stay grounded.

Among this group are, of course, students like Jennifer, who posts to glorify God, and others like Jose, a young man I meet at a public university who is very involved in his church. He never stopped talking about God or the way that Christianity informs every aspect of his life. Like Jennifer, Jose sees social media as a tool for evangelization.[3]

"I was on Facebook right before I got here," Jose says. "The vast majority of posts on my wall have to do with the church, or [are] talking about my relationship with God, things like that. . . . I'll post stuff like a really big spiritual thing for me, like let's say I had a great morning, had a great time with God."

Jose thinks that "people would automatically label [him] as religious" based on his Facebook page. For him, social media is a way to connect to people at his church but also to Christians wherever they may be. "It's a connection for me to have with people throughout the ministry, throughout, even the world," Jose says. "Friends from China and Africa and Europe who are Christians, part of the churches there. And I can invite people to events, I can talk about what is happening in my life, and even if that one person sees it and it encouraged them that day, or brought them closer to their own faith, or made them interested in seeing more, studying the Bible maybe, then I think it's worth it." Jose, too, feels pressure to inspire others through social media. He practices a particular "brand" of Christianity and posts with a particular audience in mind—his own Christian community and, even more important, potential Christian converts he's never met.

Jose actively tries to use social media "evangelistically," as he puts it, though he didn't always do so. It wasn't until he "started using it" this way and "actually it worked," he says, that he became a believer in the power of social media to draw people to Christianity. At his church, he was in charge of a "social media day" at which people were asked to use social media "as tools to share our faith and go out there and reach out to people," he tells me. They were given specific tasks: "Invite twenty people to church, share with ten people your conversion story, post up a picture of your baptism." To Jose's delight, they got some people to respond and come out to church. "I'm supposed to represent Christ wherever I go, and Facebook is no different for me," Jose explains. To him, this seems simple and obvious. Posting about religion can help bring people to God, and this makes it worth any grief Jose might get. He doesn't care if some people don't like that he's outspoken about religion. His attitude is, if you don't like his posts, then "delete me or take my post off your wall."

Evangelizing via social media is not the only thing that sets Jose apart from his peers. At no point does he mention being worried about future employers, or turning his Facebook page into a highlight reel or an online résumé. In fact, if Jose has any worry at all about the content and tenor of his Facebook page, it's that it actively and honesty reflects his commitment to honoring God. When Jose mentions having "to be careful" about what he posts during our interview, he is talking about making sure that his posts about his faith are authentic and rightly intended.

"I have to be careful, because it's like, 'Am I only posting this to be spiritual or to be religious?'" he says. "Because if I am, then that's not right, that's not fair to God, that's not being true to Him and it's also not being true to who I'm being right now .... but I do have to be careful of things like that, you know? Because again, this idea of it being my duty to use Facebook as an evangelical tool as opposed to just using it, it's natural to me. I don't ever want to appear as the false spiritual person."

Among the students I interviewed, Jose is virtually unique in the authentic way he uses social media. This isn't to say that he reveals everything about himself, or that he doesn't polish his image a bit; what sets him apart is that he appears to express genuine feelings online, which is something most young people are afraid to do. As with Jennifer, this appears to be driven by his commitment to God. Also, like Jennifer, posting with God and faith in mind seems to energize Jose rather than demoralize him.

ALIMA: YOUNG, MUSLIM, AND FEMALE ON FACEBOOK

Not all students who express their faith on social media do so in the same way as Jose and Jennifer. Some students, like Alima, use social media as a way to subvert the limitations their faith places on them.

When Alima sweeps into the interview room, she is full of smiles. She has rushed to get here and is worried she's late (she isn't). She plops down into the chair across from me and begins arranging herself, getting comfortable so we can begin. Alima wears a black headscarf that reveals a beautiful face and eyes as bright as her smile. Everything about her is beautiful—her eye makeup, which is heavy but tastefully done, her smooth skin. She was born in Calcutta, and her family moved to the United States when she was five. Her dress is conservative—she wears a long, traditional black cloak for Muslim women that reaches all the way to the floor and whose sleeves end at her wrists. She tells me that her father is very strict and "makes her" dress like this for school—which, Alima tells me while rolling her eyes, is *totally* ironic.

Alima's father left his own parents in India because they were too religious for his taste. He came to the United States to pursue his version of the American dream but also to try all the things that his family's faith had forbidden—drinking, clubbing, partying, sex outside of marriage. He got piercings and wore chains and leather jackets and had aviator shades like those in the film *Top Gun*, Alima tells me. But now? Today he's extremely religious, just like the family he left. She doesn't consider this fair to her at all—she's the one who has to pay the biggest price for her father's religious conservatism.

"Now I feel like I'm being so controlled by my father," she says. "He wants me to come to college like this"—she gestures at her clothing—"he's controlling me because he has no one else [to control]." Alima describes her father as "very pushy" when it comes to religion. "No movies, no guy friends, no friends, no social life, it's *haram*—*haram* is forbidden," she says. "This is *haram*, that is *haram*, you can't listen to music you can't do this, you can't do that." She adds, her tone frustrated and angry, "Shut up please! I have had enough of this."

Alima's mother is another story. She's the "rebellious type," Alima says with pride. She recently got her PhD, and although she follows Muslim tradition—she's a faithful wife and prays five times a day—she's not strict like Alima's father. "She's like, 'You want to have fun with friends? Go ahead. You want to go to a party? Go ahead. But know your own lines. You can have all the fun you want, but don't do anything that's going to upset God. And just don't be in relationships with boys. Have them as friends, but don't cross that line of friendship." Alima's mother tries to be the peacemaker between Alima and her father, and when Alima complains about how she has to dress for school, her mother tries to console her by saying that it's not a big deal, that it's just clothing, that Alima will always be Alima underneath, and if wearing these things is the only way her father will allow her to attend college, then she simply will have to go along with it.

Alima commutes from home to attend the public university where she is a sophomore. Living at home as opposed to in a residence hall is typical for an unmarried Muslim girl her age, but it also means that Alima's choices for college were very limited geographically—also her father's doing. She's thrilled to be going to college outside of her house at all, though, since for a couple of years she was only allowed to take online courses, which is how she earned her associate degree. Alima attributed this to "trust issues" with both parents that developed during high school, but mostly it was because of her father, of course. "I understood that they were [requiring me to attend college online] for my own safety and, you know, to keep my character and dignity," Alima says graciously. "In the beginning, I was like, 'This is so boring, I don't want to do this because online even if you do interact with people, I feel like facial expressions and eye contact are very important to a conversation, and it took me a while to get used to it, but I got used to it.'"

Alima first set up a Facebook account when she was in high school. Her mother was upset when she found out, but eventually she got over it. Alima's mother is able to access Alima's Facebook page—she has the password and sometimes goes online to check what Alima is posting, and to see what the rest of their family is up to. Neither her mother nor her father has a Facebook page of their own.

During high school, Alima says, Facebook was "drama central." At one point, another girl at her school made fun of her in a very cruel way—maybe even bullied her a little. "When I was a junior," she begins, "there was this girl who took a picture of her[self] punching a garbage can that had the beak of a toucan. Right? And she tagged me in it. She tagged me as the [toucan] garbage can because my nose is quite large." Alima points to her nose, then covers it with her hands, laughing slightly, but seeming embarrassed, too. People apparently made lots of comments about the photo—many of them at Alima's expense. "I was hurt, I was angry," she says. "I had self-esteem issues. I didn't find myself to be very attractive because of my nose. I hated my nose. But you know, on top of that people were making fun of it. Toucan Sam, Fruit Loops—you know because Fruit Loops has a toucan—Pinocchio. I was just like, 'Oh my god, is my nose really that big? It hurt, it hurt, I was really angry as well, but you know, I look back and I'm like what a waste of time. I feel like it was just a phase, everyone goes through it. But how you handle it shapes who you are."

When we turn to the subject of who Alima is online today, she tells me that if I went to her Facebook page, I'd see that her status this morning is about the rain and how much she loves it. She has lots of pictures of her family and also of cars and motorcycles. Then she tells me something interesting about her profile picture—and what it has to do with being a Muslim girl.

"My profile picture isn't of me," she begins, then backtracks. "It *is* of me, but I'm not showing my face. It's my hands full of henna." She goes on to explain that you can see her headscarf, but her hennaed hands are covering her face. Alima has very strict privacy settings with respect to who can see what on her profile, and aside from the family members she's friends with, most of the people she's connected to are girls.

The reason?

"Because I take pictures of myself without my scarf," Alima explains. "And the hair is considered beauty in my religion." Alima loves her hair, and she loves posting pictures of her hair exposed on Facebook. But only girls can see them.

"When I add a certain friend, I have them grouped by girls and then guys," she says. "If you're in the guys' group, you can't see my pictures. If you're in the girls' group, you can. [Guys will] see my pictures of the cars, motorcycles, but just not my face or hair." They can also see the profile picture of her hands blocking her face—but that is the only picture of Alima they can see. This is pretty typical Facebook protocol for Muslim girls, Alima thinks. "My Muslim girlfriends, [when] they post, they're all covered and modest. There aren't any friends of mine, Muslim friends of mine, who post pictures of themselves, you know, exposed. They always have a scarf on, they're fully covered. I mean they have makeup, but that's it. And my Muslim guy friends, a few of them like to show off their abs, but other than that they're all pretty covered as well."

Then Alima wants to talk a bit more about boys.

Alima *loves* boys and she *loves* talking about them. In her religion, she tells me, boyfriends are completely forbidden, but this doesn't stop her from having major crushes. Facebook is a great place to be in touch with boys, and Alima has struck up a very interesting deal with her mother. Technically, Alima's faith should forbid her from connecting with boys via social media. But she is allowed to friend boys if her mother approves first. In fact, if Alima finds herself wanting to friend a boy on Facebook, she and her mother sit down and have a chat about it. She'll literally go to her mother and say, "Ma, you know, I met this guy." She'll even tell her mother if she thinks he's handsome, and then she'll ask for her mother's permission to add him as a friend.

"I do that because I respect her," she says. "I know I can always add [a guy] and talk to him whenever I want, but I do that because I respect her and also because I value the trust that she has with me. Even though she can go on [because she has my password] and see who I'm adding. But I value my relationship with my mom a lot more than I did back in high school. So I ask her."

Does her mother ever say no?

She usually says yes, Alima tells me, with a laugh.

Alima finds herself wishing that the handsome boys she's friends with on Facebook could see those photos of her showing her hair, though. "But I know it's wrong, I know it's wrong," she says. "Not just in the eyes of my mother, but in the eyes of God it's wrong." Having boys as Facebook friends is difficult for a young, pretty, unmarried Muslim woman. It's tempting to flirt and to communicate far more than Alima thinks she should—something she did in high school, she admits. She doesn't do that anymore, though. "I feel like I'm strong enough to control myself," she explains. "There's always the right and the wrong, and you know the consequences of the wrong, and you know the benefit of the right. And I usually take the right decision." But Alima is not alone in this struggle. Making the right decisions around boys on Facebook is something that all young Muslim women her age face, she thinks.

Muslim girls even sext sometimes, she tells me. They talk to boys, they send pictures, and the Muslim boys like to watch porn. "I was taught that sex is supposed to be divine, something you do with somebody you sincerely love, not just, you know, a hookup around the corner," Alima explains. "I've learned if it's a bunch of boys around a phone it's either basketball or porn." Social media and smartphones make all of this behavior far too easy, in Alima's opinion, and she does not approve of it. "God forbid if I have a daughter and I see her doing that kind of stuff," she tells me. "It would break my heart, it would break my heart."

Before our interview ends, I ask Alima if there's anything that we didn't yet talk about that she'd like to discuss. It turns out, there's quite a bit.

"I'm just really afraid for the young girls, I'm afraid for the next generation," she says. "I'm afraid that they're not going to know the struggle of actually going through the encyclopedia, they're not going to have the communication skills that we need to conduct an interview or have a regular conversation with anyone on the street. I'm afraid that they're going to go through a lot of cyberbullying." She pauses a moment, thinking. "But what can we do about it?" she asks, then pauses once more. "You can't completely ban your child from using Facebook because they will find another way. You know? So, I think I'm going to have a hard time raising my kids in this country. I'm not going to be like my father, but I will definitely protect my sons and daughters from the harm that's possible because of social media. The drama involved, the amount of time it can possibly waste. The things you see that change your perspective on people. It's just sad. It's sad." Then Alima pauses one last time, and a thoughtful smile lights up her face once more, and she laughs. "But then, social media can cause a revolution, you know?"

TO FACEBOOK OR NOT TO FACEBOOK

Orthodox Jews also have strict rules about social media. But while many stay away entirely, others engage in ways that give them freedoms they would not otherwise have. One young Orthodox man I interviewed, a sophomore named Zachary, told me he is incredibly antisocial in real life, but is very active on social media expressly because "it keeps [him] in check with the secular world." "Which is very important because I go to a secular university, I live in the secular world, I don't live in a ghetto," Zachary says. "So I need to remember that no matter how I might look or how I might talk or what I may think, you still have to adjust to your surroundings a little bit."

But social media is more than a study in secular living for Zachary. It's also a sphere where he loves looking at photos of people doing all the things he's not allowed to do because he's Orthodox. Luckily for Zachary, his extended family isn't Orthodox. "What they're doing doesn't align with my beliefs," Zachary explains. "But that doesn't mean I don't want to look at pictures of them doing it. You know, I have cousins that go out once a week to this sushi restaurant, and I could never eat there. But I *love* looking at the pictures of them all sitting there in this hibachi-style restaurant. [And] if there is a party on Saturday or something, I like looking at pictures of what they've done, you know, my other friends that do things that I don't."

"I'm very grateful," he says, that social media gives him a window into the secular world.

But then I meet Dinah, a senior at her public university, who is also an Orthodox Jew but has a very different relationship to social media than Zachary. Dinah is very serious about her studies. She has beautiful blonde, wavy hair and bright blue eyes that peer out from behind her glasses. She's petite, energetic, friendly, and excited to talk about anything and everything. Like Alima, she commutes from home, where she lives with her father (her mother died some time ago) and her many siblings.

Before we start our interview, however, Dinah worries that she isn't a "good candidate" for this study. She is an "outlier" with regard to social media. I tell her this doesn't matter, and I ask her to go into greater detail about what she means. As it turns out, Dinah doesn't have any social media accounts—she doesn't even have a smartphone, which is extremely rare among the students I interviewed. Dinah does spend time online, she assures me. She watches YouTube videos of people singing, because of her interest in music. But unlike most of her Orthodox friends, social media is a no-go for her.

"Pretty much everyone I know, including and especially my Orthodox friends, have Facebook," Dinah begins. "It's very simple. If I get Facebook, I will never get offline. I'm premed, I have things I have to do. . . . If I got it I would never stop. I will get addicted to Facebook, and I'll never get off, and I'll spend all my time in the house." Dinah backtracks a moment to clarify her comment about how her Orthodox friends "especially" are on Facebook. "Most Orthodox people I know at this point have Facebook. The Hasidic people don't . . . . and they have their reasons, which are usually about not running into things you don't want to see." She laughs and shakes her head. "It's funny, because for a community that was so anti-Internet originally, and still is in certain pockets, [Orthodox Jews] have nothing really wrong with the Internet. They have an issue with pornography, and the ease with which you could escape from Judaism, escape from being an Orthodox person, because you see other people and you see how they live."

Dinah's Orthodox friends who are on Facebook struggle with it. Comparing their lives with the lives of those who don't have such limitations can be extremely difficult. "I have one friend who is incredibly desperate to get out of her parents' house," Dinah says. "She hates living with them. She doesn't hate her siblings, but she doesn't really get along with them. . . . And when she sees the posts of the other people who had internships with her on Facebook and she realizes that they're living in college dorms and not their parents' house, that they are not as tied as she is, I think she gets very upset about it. For the past four years I have watched her come to hate her circumstances more possibly because she's seen others. I mean, she's met these people, she knows that they're different than her. I think that constantly being reminded isn't so easy for her."

Whereas the way that Facebook showcases the secular lives of Zachary's family members offers him a much-desired window into what he's not allowed to experience—a window he values—Dinah's friend only experiences pain because of what she sees. The phenomenon of comparing oneself to others is particularly acute when a young Orthodox woman is constantly perusing the lives of non-Orthodox friends because the distance between what she's allowed to do and what they can do is all too evident. The window Facebook offers only makes her circumstances seem more unbearable.

Then I interview Ephraim, who is also an Orthodox Jew, and he, too, warns me that he is a "bad specimen" for this study. I tell him there is no such thing, and off we go.

Ephraim grins nearly constantly and cracks lots of jokes. He tells me he looked me up before coming here (one of the only participants who admits this) and knows I've written a lot about sex. His eyes twinkle, and he smiles knowingly. Then he asks if we're going to spend time talking about hooking up.

"No," I tell him, laughing and shaking my head. "Sorry to disappoint you."

Ephraim almost went to a university whose student body consists nearly exclusively of Orthodox Jews, but he's happy he chose this coed public university, where he gets to mix with people of different backgrounds. He likes going to classes with women, too, and at one point he marvels at how different this reality is from that of his home life—even the basic fact that he is sitting in a room with a woman now, just the two of us, doing this interview. Ephraim even lived away from home for a while, in a "basement full of guys," he tells me, all of them Orthodox. "It was wild," he says, gleeful at the memory.

Ephraim is clearly very pleased that he has done things differently than his Orthodox Jewish peers. "I define myself a lot by otherness," he says. "By the fact that I am able to perceive myself as unique in the place where I come from. . . . I was always somehow different than everybody else."

Like Dinah, Ephraim avoids social media.

Despite starting off our conversation by saying he might be a "bad specimen" for my research, it turns out Ephraim cannot wait to tell me all about why he's so different than everyone else. He informs me that our interview will probably be interesting precisely because of this, and it will show me "another side" of things I haven't yet heard. Opposition to social media is a central part of Ephraim's self-understanding. His stance against it is crucial to his identity and how he sees himself in relation to others—how he sets himself apart. "It became part of my persona," he says. "Like, I'm a non-Facebook guy. Which is also part of the myth . . . .," he adds, trailing off.

As in, the myth of being Ephraim.

"In terms of my perception of myself, [having Facebook] would make me more like everybody else, and I don't want to be like that. But this is me admitting to my own fantasy," Ephraim tells me. He also admits he's afraid that if he tries out social media, he'll like it so much he'll get addicted to it, and that would really undermine his sense of self.

"I hate social media," he says, laughing, referring to the mere presence of it, the temptation it poses both to himself and to everyone around him. "I despise it. I think, first of all, that the Internet as a whole is a terrible distraction. I think technology as a whole is a terrible distraction. I am one of those Luddites, you know, the ones that would basically be, I would pretty much be happy in the Stone Age, you know? Horse and buggy works fine with me. I don't think it's added anything to what's essential to life. I think it's just sped it up and added quantitatively to what life is. And, slowly, it's adding qualitatively to what life is, and I think that's bad."

I ask Ephraim what he means by this. "[Social media] is sort of usurping real knowledge, real study, real concentration," he tells me. "All those things that make for greatness, that make for, you know, hard work in the intellectual sense. It's making it basically obsolete."

Unlike Dinah, who finds that her Orthodox Jewish community is very open to social media, Ephraim feels that in his community, "there is a strong anti–social media movement, if you can call it that," he says. "It's almost universal, so it's not really a movement—they all hate social media. They're afraid of it. I think any religious community is, especially one that's as tight and close-knit as the Orthodox Jewish community. So, yeah, I think it sort of protects me by default." Ephraim's nonreligious friends are all on social media, of course, and he says they've "begged" him to get on Facebook so they can stay in touch more easily, but he always refuses. "I think another part of why I don't like social media is the facade that you have to [create] . . . .you have to create yourself in a virtual medium, and I hate that. I just have a knee-jerk revulsion of it, of this having to create your persona in this very superficial way. Putting up pictures and doing this, people commenting, it's just so silly to me. And I'm just not interested. I'm not buying it."

Both of Ephraim's parents have "bought it," though—they have Facebook accounts, and they wouldn't mind if Ephraim got one. But, interestingly, while Ephraim's parents and he can have Facebook accounts, it's a different story for his sisters. There's a clear double standard. "Because girls are considered a lot more vulnerable," he explains. "I don't want to generalize, you know, but my sisters, they're not even allowed to have phones with texts. Everybody's too nervous about protecting our virgin women." Ephraim says this last bit with a smirk on his face followed up by a roll of his eyes.

Ephraim and his Orthodox Jewish guy friends are also allowed smartphones, and one of the consequences is that they can explore sex in ways that girls are barred from doing. He starts telling me about how, for a while, they were all obsessed with ChatRoulette—a Skype-type platform that connects people randomly via video—you never know who you are going to get. I wonder if going on ChatRoulette is an example of the "wild times" Ephraim told me about earlier on in our interview, when he mentioned living in that basement with lots of other guys.

Ephraim refers to it as part of "the underbelly" of the online world.

"My friends introduced me to this, and obviously it was about getting girls, you know?" he says, laughing. "You see a fair amount of naked guys on it, you'll click to get out of it, like, 'He's naked, *next.*' But there are also a lot of girls on there," he adds, giving me a knowing look. "There would be these long marathons and we would be on three separate computers, different ChatRouletting. It's a fascinating underworld universe, and I think it's very dark. And I know friends who have gotten girls to strip for them and things like that. It just shows this desperate need to get out of your own universe and to live in a social universe that's more accessible. And I think online offers that."

Unlike the devout Christian students I met, who worried so deeply about honoring God in their posts or staying away from social media if it tempted them away from God, Ephraim clearly loves his memory of the era when he and his friends had these online adventures, more so even than Zachary. But he also hates the way they became obsessed with it. Eventually, "I stopped it, and I did sort of force my friends to stop it," he says.

Internet, smartphone, and social media addiction is a regular theme for Ephraim. He talks a lot about how he has an addictive personality. He refers often to his fear that if he tried social media, he'd become addicted to it, which is why he refrains. He feels addicted to his smartphone, for instance. Then again, while at regular dinnertimes he finds it difficult to put his phone aside, Ephraim doesn't struggle with unplugging for Shabbos each week. "It's not difficult because you know that it's not even an option," he says. "And there are so many other things that we do regularly that you don't do on the Sabbath, it's just one of those things."

Ephraim worries that "we're becoming more online people than we are people with just an online outlet." This is tragic, he believes, but he credits his religion with protecting him from the worst of it. "I think it's disrupting to a lot of people and the way they socialize," he says. "A lot of the people I know happen to not be such technology-crazy people because of the religious thing, and I feel like I don't suffer as much as others would. I think if I hung around more with more secular people, I would feel it more acutely."

FREEDOM *FROM* EXPRESSION?

For Dinah and Ephraim, the choice to stay away from social media itself builds character. And while the Christian students sense that God is watching, this seems to inspire them toward a kind of positive zeal in their faith. Those devout students who are on social media find it to be an incredibly useful, rich, diverse, and, at times, subversive, playful tool for religious expression. That so many college students—because of image-consciousness and their fears about the judgment of future employers (and the judgment of just about everyone else as well)—have taken religious and political expression off the table for their online lives should at least give us pause.

What does it mean that social media is not an inviting space for many students to be honest about their faith? Or their politics, for that matter? Is it a significant loss that young adults need to be so careful in these areas? And what does it mean that colleges and universities themselves—institutions that boast about turning out good, activist, tolerant citizens of the world—are also training their students that, above all, they must be concerned about what everyone else will think, especially future employers, and "post and delete accordingly" on social media? Are there ways that we can shift the conversation such that political and religious views might not jeopardize your ability to get hired in the future? Or is it simply naive to think that might be possible?

What is clear is that young adults long for places where they can be themselves without censorship or fear of repercussions. To find such freedom, they often turn to anonymous platforms, which come with their own set of problems.

6

VIRTUAL PLAYGROUNDS

The Rise of Yik Yak, the Joys of Snapchat, and Why Anonymity Is Just So Liberating

I do follow Yik Yak, kind of like a bad soap opera.
                    Gina, junior, private-secular university

It's a little liberating to be able to post what I really think without any fear of being judged for whatever reason. It does, however, disturb me when I read racist or homophobic or generally hateful comments. I used to use Yik Yak almost every day until I got overwhelmed by the racism that exists on my campus.
                    Hope, first-year, private-secular university

GRACE: MY TRUEST SELF

When Grace and I sit down for our interview on a bright, sunny weekend morning, she says something I almost never heard other students say: "I actually think I'm arguably more authentic online." Grace is of Chinese descent—petite, with long, shiny black hair and a friendly demeanor. By the time we meet, I've grown so accustomed to students talking about how they can show only one side of themselves—the happy, positive, even inspiring one—that I'm a bit taken aback by what Grace says. How has she found a way to be *more* authentic online? How has she (*has* she?) escaped the pressure to appear happy?

The answer is simple: total anonymity.

Grace is most active on LiveJournal, a blog site where people can write to, interact with, and follow others. She *loves* LiveJournal because nobody on the site knows who she really is. Well, *almost* nobody.

"It's basically a bunch of communities, you can join them, you can talk about things that interest you," she says, smiling enthusiastically. "So there's one community that is for a music group that I enjoy, and I've made a lot of friends on that website, but they're not people that I know in real life. Well, there are a few I know in real life, but mostly, like, strangers." Then Grace captures the blessing and the curse of anonymous platforms: "You can be anonymous basically, so you can say and be who you are, and no one can really judge you for it because they don't know who you actually are. And because you have that mask to hide behind, that, I think, enables a lot of people to be very authentic and very genuine in expressing what they think. Sometimes *too* authentic. Sometimes people will say things that are mean, and if they were in real life they would never say something that mean, but because you have this shield, this mask of [anonymity]."

Grace uses a pseudonym on LiveJournal, which means she's not risking her "brand" or courting disaster with future employers, as she would be on Facebook. If anyone Googles her, they're not going to find her LiveJournal account. And that is very liberating.

"No one uses their real name, and it's not connected to Facebook," Grace explains. "It's not connected to any other platforms unless you choose to, and very few people choose to, so a lot of the people are just some sort of [a] user name and that's who you talk to. And, again, it feels like a very trusting place, and that's why it's easier to speak up."

As we talk further, it becomes clear not only that Grace is different on LiveJournal than on other social media sites but also that she's different there than she is offline. In real life Grace is not very outgoing, and she has only a small group of very close friends. As she tells me about her LiveJournal self, she's almost gushing.

"I think [when I'm on LiveJournal] is when I feel like I'm being truest to myself," Grace explains. "Because I'm able to type how I feel, and it's just more comfortable that way, for me at least. Well, *one* because I have a little more time to think about what I want to say, and *two* because if I say something that I didn't mean, I can delete it before I send it. I can say it and then correct myself and say, 'You know, maybe on second thought, that's not what I actually mean. This is what I actually mean,' and then I can say that instead." Scholars who study social media refer repeatedly to the control people want, feel, and assert over their online communications, one that often comes at a detriment to their real lives.[1] Grace's comments highlight this desire and also show how heightened her sense of control becomes when the platform is anonymous. The anonymous platform not only offers freedom in an online world that is largely about maintaining a happy appearance in the service of pleasing one's audience but also acts as an outlet, allowing Grace to escape from *real* life, too.

Grace has "friends" on LiveJournal, but they are all anonymous, too. "So it is *like* having real friends, because they're there for you and you're there for them," she says. "I'm able to just speak with them very candidly." Grace loves that she can be "emotional" on LiveJournal—another thing I almost never hear from students about their lives on social media. Almost everyone thinks that emotions other than pure bliss are off limits. In the online survey, I asked students to respond to the statement:

I am open about my emotions on social media.

Only 19 percent of students agreed with this statement. Nearly all the rest (74 percent) said no. So Grace is quite rare in seeing social media as a space for her emotions, especially because she means all kinds of emotions, not just positive ones. It was so common to hear students talking about forced positivity that I began to wonder whether there were any college students who do not follow these rules in their social media lives.

In Grace's case, anonymity allows her to be *more* honest and *more* herself than she is in her real life. She can be someone totally different from the person her friends, family, and teachers know. "I think, [who I am on LiveJournal] is not just a part . . . ." She pauses here, then backtracks. "It is *just who I am*. But who I am is expressing something in this media. Through this format, I'm expressing myself. . . . There are many forms of me, and I think that social media is just another way that I can be me, and interact with more people than I can when I'm here. I don't feel like it's an extension or a different part of me at all. It's just who I am, expressed differently."

Then, suddenly, all the expression of enthusiasm about Grace's online life comes to a screeching halt. It happens when I ask her if she is also like this on Facebook.

First, her face falls. Then, Grace shakes her head. "I hesitate to post everything that I would like to talk about," she begins, and the young woman who just spoke with such joy about the incredible self-expression she finds on LiveJournal is gone, replaced by someone resentful and frustrated. "I'm not someone who posts statuses every day. I'm not someone who shares a bunch of things. That's mostly because on Facebook there's a different dynamic that doesn't exist elsewhere." Grace goes on to list the many things she does not like about the "Facebook dynamic," which include "the concept of 'liking,'" the comments feature, that it's all about who's popular and who's not, that people post things specifically to *get* "likes" so "they'll feel good about themselves." "It is a popularity contest," Grace says, and she has no desire to compete with others like this. She finds it exhausting, demoralizing, and *inauthentic*. Worse still, Facebook is attached to your real name. "It's very public, and I don't want everyone to know what I'm thinking or saying all the time," Grace explains. "I don't like that anyone that is a friend of a friend can come to my page and see what I've been up to. It's none of their business."

For Grace, even the potentially terrible disadvantage of online anonymity (bullying, cruelty, meanness) is worth the risk for the satisfaction she gets out of getting to be who she really is, "expressed differently" as she put it—a satisfaction Grace feels she'll never be afforded on a platform like Facebook. Anyone who reads the newspaper knows about this downside to anonymity on social media. "People can be whatever they want, and if they get banned, they can just make another account," Grace says. "There is that ability to discard an identity and just come back to it later or create a new one and keep going regardless of what happened to the other one. So I definitely think people are more loose with their tongues online. . . . But that's the nature of having that screen to hide behind. That's the nature of online socializing."

Here, Grace is specifically referring to sites where people have usernames that are fake and are not associated with their real identities—which is what allows them to "discard an identity," as Grace puts it, and simply create a new one to replace it. Grace has chosen to live with this side effect of anonymity and seems to have made her peace with it. But the very anonymity that allows Grace to feel so empowered, to be as expressive about herself as she wants, is also what can make certain platforms havens for bullying, racism, sexism, and other forms of discrimination and threatening speech—speech and also behavior that are nearly impossible to control, and that cause the young adults who are its target to suffer deeply. As Grace points out, users can simply shut down one account and create another to continue their maliciousness, taking advantage of and hiding behind the anonymity the platform provides.

I push Grace a bit, on whether she thinks this ability to discard identities is a good or a bad thing. She doesn't think "there's a judgment call for that," so it's neither good nor bad—it just part of the deal. "It's not something that we can really change, and it's not something that people are willing to give up," Grace tells me. "This freedom of being able to be whoever you want to be, and in this case, be multiple people . . . . personally, I think that's a little bit dishonest, but there's nothing stopping you from doing that."

I ask Grace if people who "discard an identity" and create a new one can do it in a positive way. She shakes her head again and sighs. "You could, but most people don't do that," she says. "Most people only do it because, the only reason you would need two identities is if you're not satisfied with the first. So, if you're a really positive identity, it's very unlikely that you would try to create another one doing more positive. You would just build upon the one that you have. Whereas if you've built a very negative identity that's received negative criticism, then it's easy to say, post on that account, 'I'm going to shut down. I'm leaving,' and then you can turn off your account and no one will bother you ever again, while you pull up another window and make another account and just keep going. It's convenient for people who don't like what they've been doing."

Grace clearly doesn't like this situation, but freedom, both good and bad, comes only with anonymity. And despite the drawbacks, Grace's anonymous online life is hugely positive for her self-esteem. She enjoys it, thrives on it, feels liberated by it, and finds much-needed community through it. Given the demand for positivity online, the professionalization of social media, and the branding of the self, students need to seek out other ways to be themselves online. Anonymity allows them to do that.

LET YOU BE *YOU*

Grace is by no means the only student who spoke at length of the joys of anonymity. Susan, an English major, told me how much she loves Tumblr, where she uses a fake name. On Tumblr, Susan says, "I can speak out more and mention things that I'm really interested in that, for example, maybe if I said out loud in public, people would think I'm strange." When I ask Susan to give me an example, she mentions not some deep, dark secret but her unbridled enthusiasm for Disney princess movies. Susan doesn't want people making fun of her for this, and Tumblr is a safe place for her to post about it.

"It's not that I portray myself [on Tumblr] as a different person or anything," Susan says. "For example, my Facebook, it has my name on it. Everyone on there knows me, so my thoughts would be similar to what I would say, in a normal public atmosphere, but my Tumblr or any other account my friends don't know about, I'm more open on there." Susan feels her Tumblr username acts like a shield that protects her and allows her to say certain things "out loud" that she would never say either in person or online in places where her real name appears.

Then, during a visit to an evangelical Christian college in the Southeast, I meet Angela, a tall, curvy young woman with long blond hair who smiles a lot and has a sweet, soft voice. A glint appears in her eyes, as if she has a delicious secret, and soon I find out Angela *does* have a secret—well, she used to—a secret Twitter account. She made it when she was feeling down and abandoned by her friends. And, boy, did she love it.

"I do have a Twitter account that's anonymous," she tells me with a laugh. "I made it—this is going to sound stupid—I made it to get followers, because I thought it'd be fun. I was bored, it was after some [bad] stuff happened in freshman year in college, I only had a couple of friends." Angela had been paying attention to people on Twitter for a while, seeing how the site works ("what gets the most retweets, what gets the most 'favorites'"), and she thought to herself, "I could do the same thing." So, she tried it. "It started out as an experiment, and then, speaking my mind if I was frustrated with my friends," she says. "I would vent on that account, and no one would know it was me. I think that's why I did it." Angela started the account when she was feeling alone and rejected by her friends, so posting to it and gaining followers gave her a sense of social connection that helped counter some of the disconnection she was experiencing in her real life. The anonymity of it added a layer of safety and a feeling of protection for Angela.

Angela hasn't posted to the account in about six months, but when she was on it, she especially liked the freedom that anonymity granted her to speak her mind. During this experiment she learned that posts with hashtags get the most retweets and "favorites." And it was very important that the tweets not be personal. "If you were actually using it as a personal account, people didn't care," Angela says. "It was the superficial, it was almost like a *celebrity* account, if that makes sense. So it was, like, putting something out there that people want to read. They did not want to know the person behind the account, if that makes sense." This, for Angela, added yet another layer of safety—not only were her posts anonymous, but Angela wasn't posting on things that were dear to her, so even if people ignored her tweets or said something nasty, she could better shrug off the response (or the lack of one).

Eventually, Angela amassed thousands of followers.

How did she do it? "You just start interacting with different people, different groups," she tells me. Actually, Angela and Susan have something in common. "I really like Disney, so one of the things of mine, like, my background was a Cinderella theme, so you get in with some of the other Disney people, they all retweet you. It's the retweets, that's how you would get follows." Angela says she "made it to vent," but eventually she got to a point where she was just "playing with it" to see how it all worked. Getting "favorites," retweets, and new followers got Angela really excited. "You would feel successful, I guess," she says. "It's like getting points in a game."

Interestingly, unlike Grace and Susan, Angela says she posted many of the same things on the accounts attached to her name as she did on her anonymous Twitter account. "The only thing that would different would be if I was really angry, I could put some cuss words in my anonymous one," Angela explains. By posting more or less the same things to all her accounts, Angela learned something interesting. "People are more willing to interact with you if they don't know the person behind it," she says. "It gives you anonymity, but it gives them anonymity too. . . . I think it's less intimidating [than Facebook]. Whereas, if it's your name on there, [people] might not necessarily be as willing to retweet things. So, my name on that [Twitter] account was not a person's name." Occasionally, Angela would get more of a response to a post on Facebook, but it really depended on the subject. On Facebook, people respond more to personal things.

Angela rarely uses social media at all anymore, except for saying "Happy birthday" to people and keeping in touch with family. The Twitter account was a fun experiment, though, and she enjoyed it while it lasted. Only a few friends and Angela's boyfriend knew about it. Well, and Angela's younger brother. It's here in our conversation that Angela shows just how detached she has become from it all. Angela's brother had his own Twitter account, and when he found out his sister had so many followers on Twitter, he asked her "to give him a shout-out" to help get him more followers.

Instead, she just gave her account away. As she tells it, "I said, 'You can just take this one.'"

THE JOYS OF THE "DISAPPEARING POST"

While it's true that just about everybody seems to be on Facebook, that doesn't mean that it is the students' favorite platform. It's more like a necessary evil. Because everyone is on Facebook, even professors, it's useful for staying in contact with people and communicating about class projects. And, of course, Facebook functions as a social media résumé. But it is no longer as much fun for students as when they first got their accounts. Even Instagram, which is infinitely more appealing to students than Facebook, can be stressful because it's generally attached to your name.

Snapchat is a very different story. College students *love* Snapchat.

They love it because their "snaps," or posts, disappear within seconds. You take a photo, give it a caption, and send it off as a visual text to a friend or group of friends. Yes, your profile is (usually) associated with your name and your real identity, and everyone knows that people can take screen shots of your "snaps," but as long as you're not reckless with what you send and who you send it to, you can feel pretty safe that it's really going to go away just as the app promises.

Students express nearly universal adoration of Snapchat. As people explained why they love this app so much, it became clear that college students are longing for a space where they can be themselves and not have to worry that it will come back to haunt them in the future, when potential employers start looking at their online activity. They can rest easy knowing that their Snapchat photos, selfies, and comments will fade away.

The very fact that snaps disappear has led Snapchat to become known as the "sexting app."[2] And, yes, people use it for that. But for the college students I spoke with, this is an impoverished and limited understanding of Snapchat's true delights.

College students can be *silly* on Snapchat. They can be ridiculous. They can say dumb things. They can take goofy, ugly, unbecoming photographs and show them to other people. They can be sad, they can be negative, they can be angry, they can even be mean. They can be as emotional as they really feel. They can be *honest*. And it's true, on Snapchat college students feel they can be sexy. But most of all, they *play* on Snapchat and they engage in all kinds of foolishness. And that's why they love it.

On Snapchat college students feel they can do all the things they've learned they're not allowed to do on Facebook or any other platform that is more "permanent" and attached to their names.

Matthew tried to explain the difference between Facebook and Snapchat to me, and why Snapchat is much more fun. Like many students, Matthew goes onto Facebook a lot, but not to post—posting is too time-consuming and too much work, because every post has become so high-stakes. Mostly, Matthew just scrolls through the feed and lurks, checking out other people's updates and photos. But Matthew loves Snapchat and goes on it all the time, and unlike with Facebook, on Matthew actually participates.

"When I'm bored," Matthew says, "I'll snap a picture of something random, send it to, like, five people and wait for somebody to respond. [Snapchat] is really simple and fast, and it's a way I can see what all my friends are up to, especially all my friends back home, all over the state and stuff. They'll send me back a snap of them like, I don't know, doing whatever, and I'll be like, 'Oh cool.' It's almost like I'm there, so it's a way to share experiences from far away, and it's so quick and simple."

Snapchat offers its users a way to connect without judgment, a possibility that college students, who feel they are constantly being watched and evaluated, long for. That's what Facebook is supposed to be, too, in theory, and perhaps it was at some point. But for Matthew, Facebook is almost exclusively a place to wish someone happy birthday, to celebrate engagements and marriages or comment on family matters. The other thing Facebook is for is announcing "something huge" that happens in Matthew's life—so he can cultivate his highlight reel. "Graduations or, you know, success in baseball," Matthew says, and pauses as he searches for another example of what might prompt him to post something on Facebook. "So, say I threw a no-hitter, I would probably take a picture of the scoreboard with all the zeroes across it, and I'd post it on Facebook like, 'My first no-hitter in college.'"

But then Matthew tells me he'd "snap it" to people as well, the people, in this case, being his friends—his *real* friends, not his pretend Facebook "friends." He might "snap" his baseball no-hitter, or he might "snap" a picture of himself nearly asleep in class and write along with the photo, "Crazy bored right now" and send it to "everybody, so hopefully they'll send something back and help entertain me," he adds with a laugh. "There's no effort in Snapchat. I just look at something, hold up my phone, and hit a button, then I send it to everybody. . . . Snapchat has [My] Story for 24 hours, so for 24 hours, you can snap a bunch of stuff, put it on [My] Story, and at any time people can look and see what you've been doing the last day or so."

That lack of effort Matthew mentions is another part of Snapchat's appeal—especially when "crafting" just the right post these days sometimes takes *forever*. People have to worry about so many things when getting ready to say something online: will people "like" it, will they ignore it, will it turn out to be a comment later regretted, will it offend anyone, will it worry anyone, will it show me in a positive and happy light, will it display my successes, will it make me seem enviable—this list could go on and on. It's very difficult to find a student who makes regular updates on Facebook today because deciding what to post and wording and editing it takes so long and then you have to contend with all the stress that follows the posting. Updating social media—the kind that sticks around—can be exhausting and feel like a job.

With Snapchat you can just relax and play around. Another thing that's nice about Snapchat is that it's one of the only platforms where you don't have to worry about cultivating an image, Matthew tells me. You can snap pictures when you're drinking alcohol, or simply not show your "perfect" side. This is a huge relief for him, and another reason he loves it so much. He explains, "My image on Facebook and Twitter is a lot better than it is to my friends on Snapchat."

By "better," Matthew means more airbrushed and less himself. Snapchat brings Matthew so much joy because he can just be who he is, however he is at the moment, even if it's not happy or particularly becoming. It's a platform for being "real."

So many students I interviewed told me more or less the same things as Matthew did about Snapchat. After hearing one student after another talk about the importance of always appearing happy and positive, and the concern about possible repercussions if they do not, it became clear that there was a connection between the pressures of Facebook and its waning popularity among students. There seems to be a direct correlation between the professionalization of Facebook and the "branding" of one's name on certain social media platforms and the rise in popularity of apps like Snapchat and Yik Yak. That Snapchat promises a kind of "no repercussions" experience is absolutely beguiling to everyone who feels burdened by the pressure to appear perfect everywhere else.

Snapchat, I think, provides a kind of catharsis. It's the place where everyone lets off steam and celebrates imperfection. Of course, the speed with app rises and fall in popularity today is so meteoric, by the time this book comes out Snapchat may have already faded into oblivion. But I have no doubt that the "professionalization" of social media is only going to get more intense as young adults learn to toe the line at even earlier ages. As a result, young adults will keep looking for places that allow them to be silly, to be ridiculous, and to have a little fun on social media, whether it's Snapchat or some other app that allows disappearing posts or anonymity. Only by ditching their real names, it seems, can young people truly be themselves online.

Of course, as indicated earlier by both Grace and Emma, there can be a dark side to anonymity, too.

YIK YAK: A TOTAL TRAIN WRECK

That dark side is called Yik Yak. I first heard about it from a sophomore at a northeastern Catholic university.

"Have you heard of Yik Yak?" she asked me.

When I answered no, she explained, "I don't know if it's new, but I recently discovered it." She continued, "People can post anonymously about anything, and whoever is within your area can see it. So going on and seeing what's going on here . . . .people will post like, 'Yeah, who's down to come over and have sex or whatever?' and 'Who's down to party? Who's down to smoke?' I feel like if that's what's presented online, I'm *not at all* wanting to be a part of that."

Yik Yak hit college campuses like a lightning bolt. When I started doing interviews for this project, not a single student mentioned Yik Yak because it didn't yet exist. It popped up over the summer before students went back to school, and it became ubiquitous everywhere I went, seemingly overnight. Suddenly, everyone wanted to tell me about Yik Yak.

Yik Yak is often described as an "anonymous Twitter," and each campus has its own Yik Yak (more or less). Yik Yak functions like another infamous app (Tinder) by using the GPS on your phone to locate you. To log on to a certain college's Yik Yak, you must be on or near campus. It also functions a bit like Reddit, in that people "upvote" and "downvote" whatever gets posted, causing the most popular posts to stay at the top of the feed where everyone can see them, while the least popular ones drop so far down they basically fall into oblivion.

Yik Yak's arrival has sparked a media frenzy, with splashy articles in *USA Today, Slate,* and the *New York Times* and stories on CNN.[3] It has quickly become associated with scandal and hate speech as well, as in the case of the students at the University of Missouri who posted racist and violent threats and were identified through their IP addresses and charged by local police, confirming for all that Yik Yak remains anonymous only if its users stay within certain boundaries.[4] What those boundaries are, exactly, and whether they include racist, sexist, and other hate speech is still evolving.

But the best way to understand its presence, its use, and its effect on campus life and culture is to hear students themselves talk about Yik Yak. Many of the women and men I spoke with claimed to be voyeurs on Yik Yak—it's a train wreck, and they can't tear themselves away. But I did meet a few people who admitted to not just lurking on Yik Yak but participating, too.

One such young man, Justin, a junior at a private southwestern university, brings it up early on in our interview, after I ask whether he thinks people are honest when they post on social media.

"So, for example, Yik Yak is a social media where you're entirely anonymous and I think [people] are authentic oftentimes in ways that you wouldn't want them to be," he answers. "When you're entirely masked by anonymity, you can say exactly what you think. I think on something like Facebook, the temptation is to present yourself in the way that you want someone to see. So that can be both authentic and a little bit disingenuous. I mean, you're authentically telling people what you want them to think you are, but you may or may not actually align to the image that you present."

Many students find the anonymity of Yik Yak thrilling. But in a social media culture where everyone has grown so careful, so image-conscious, and "a little bit disingenuous," Yik Yak can also be terrifying. Anonymity is liberating, but sometimes it unleashes impulses that might better be kept in check.

"Oh, I do post," Justin says. "For me, Yik Yak is kind of a game of trying to get bigger 'upvotes.'" Like Rob, who lives for "likes," Justin enjoys the affirmation he gets on Yik Yak even though it's anonymous. "Even though I don't necessarily get the credit for it, I get the self-validation," he explains. "I mean, *I* know I posted that 'yak' and then it got 'upvoted.' But I also just really like seeing it. . . . It really lets me peek into other people's lives because they tell you *exactly* what they're doing at that moment on Yik Yak. So it gives you a feel for, I guess, what the 'kids on campus' [making air quotes] are doing."

Even though Justin posts on Yik Yak, he sees himself as different from the majority of users. He doesn't feel implicated by what gets discussed because he doesn't see himself as a primary participant—he just watches everyone else talk about it and chimes in here and there.

"They confirm every stereotype," he says. "They'll literally tweet, 'I just had sex.' A *graphic* description of what kind of sex they're having. They'll talk about drugs. They'll talk about, I mean, it just spans the gamut. Not everything is completely wild and inappropriate, but it's interesting to see that they're willing to do that. And obviously sometimes you read that and you're like, 'I think you're making that up,' because they know that they can tweet whatever, or 'yak' whatever, and no one knows." It's interesting to note that Justin uses "they" when discussing what he sees on Yik Yak, as though the people who comprise this "they" are distant and disconnected from him and not his very own peers, and as if Justin doesn't participate in all of the commentary, when he regularly does.

Justin then gives me examples of a couple of his more successful "yaks," which involved comments about the stupidity of the football players on campus, and how rich his fellow students are. "I wouldn't say it if I thought my name was right next to it, but [these were] just observations that I had that I knew I could get away with saying and everyone thought that was funny so they upvoted them," Justin says. "I mean, [what I said] is not extremely offensive. It's just one of those universal statements that's interesting." Justin, in other words, has a complicated relationship with Yik Yak's anonymity. When using Yik Yak, he is more free than on other social media platforms, but he's also playing to the crowd.

And sometimes that crowd resembles a mass audience. "Yik Yak would definitely reinforce the movie stereotype of college because that tends to be what I see on there," Justin says. It depicts *Animal House* behavior, I heard on several occasions. Some people said that Yik Yak would be the "death of Greek life" on campus, because of both the way Greeks get torn apart on it by non-Greeks and the way the Greeks themselves tear each other apart on it.

In the online survey, 272 students chose to answer an essay question about whether they think the opportunity to comment anonymously on apps like Yik Yak is something they desire; about 28 percent of those who responded felt these sites were an important opportunity to be honest and, at times, bitingly so. Many students stressed that anonymity is important—they crave it, love it, consider it a positive thing, debauchery and all. They felt these sites were fun, addictive, and at times silly (in a good way), as well as liberating, freeing, creative, and important in an era in which online privacy seems not to exist. Another 21 percent felt "anonymity is a mixed bag," mentioning the positive aspects but going on to discuss how anonymity leads to bullying, racism, and sexism and reveals some terrible things about your peers. *Nobody* likes this aspect of it. There is no doubt that even students who are adamantly against such sites as Yik Yak, and against anonymity on social media in general still can't tear themselves away, and are lurking, reading, commenting, and using these apps in great numbers

And many of the Greeks fear, in *too* great a number.

Mark, a junior and a member of a fraternity at a Midwestern public university, says that some of his brothers have called Yik Yak a "cancer" on Greek life. When Mark first mentions Yik Yak, though, it's with the typical "I can't pull myself away" attitude I see among so many students. It comes up because Mark is talking about all the "drama" that happens on social media. "I personally look at Yik Yak as more of a comic relief, just seeing people go back and forth," he says. "There's definitely issues that are kind of immature. . . . It's kind of stupid but funny at the same time."

For Mark, the entertainment value is too great not to go on Yik Yak, but there's also something "unstoppable" about it. The anonymity allows people to take things too far. "Sometimes you can only do so much to stop it. [With] Yik Yak, you can't tell somebody to stop it because you don't know who that person is. But on Facebook, where you can obviously see their name, or Twitter, you can single that person out or send them something like, 'Hey, don't, don't do that.'" Mark pauses a moment, then seems to feel guilty about being on Yik Yak at all. "Just, honestly, I probably shouldn't have it. . . . I mean, I downloaded it, I don't know why, but I just don't understand the thing behind it. I think it's a faceless Twitter so you can tweet without people knowing who it is. But, I don't know how to explain it, it just seems so bizarre, and Yik Yak really dampers things. It's basically, I think of it as a place to trash-talk other fraternities and sororities."

Though Mark initially referred to Yik Yak as "comic relief," he quickly shifts to how the anonymity makes it hard (even impossible) to control what goes up, then to the guilt he feels about being involved in it: "That [trash talk] really happens a lot. . . . Earlier in the year there was a lot of bashing between Greeks and then it kind of died down a little bit, but I feel like it's just to bash people. It'd be interesting to hear the guy who developed Yik Yak, and just see what he said." Interestingly enough,

Yik Yak was created by two frat boys from Furman University in South Carolina. Perhaps it's not so surprising, then, how people operate on Yik Yak, since the offensive behavior and the debauchery seem to go hand in hand with stereotypical college fraternity culture.

There is a backlash against Yik Yak by the Greeks on Mark's campus because of all this "bashing" between different houses and the way frats and sororities use Yik Yak to start rumors about each other.

Yik Yak, a number of students tell me, is used to spread lies on campus.

And fear. At one school I visited, a student announced on Yik Yak that he was going to "shoot up the school." Other students who saw this went to campus security and the administration to let them know about the posts. Other students started responding as well, posting that to say such things on Yik Yak, "isn't cool." "It was crazy," one student told me about the incident. "We get alerts to let us know what is going on, on campus, so it started sending stuff, and people were going crazy." The posts about the shooting kept rising to the top of the feed, and because of the anonymity of Yik Yak, no one knew who was behind them—at least not at first.

But, of course, if the police get involved, exceptions can be made, IP addresses can be traced, and Yik Yak doesn't have to stay anonymous. The person who had made the posts was eventually identified and caught. "You think of [Yik Yak] as anonymous," a student commented. "But they can find out who you are."

Now add slut-shaming to all of the above problems with Yik Yak.

There is *a lot* of slut-shaming on Yik Yak, students say. A horrific amount. It is sometimes used to call out women on campus by name— using names is supposed to be forbidden on Yik Yak, but people still do it, or they use clever ways to identify a person without using the actual name. Much of what is found on Yik Yak is also incredibly racist. During the interviews, students described Yik Yak as a sphere where you go to see the dark thoughts and feelings of the student body revealed, and they marveled at how shocking it is to see exactly how awful people on their campus can be. They also remarked on how there is plenty of silliness on Yik Yak—humor and fun, too—but the side of Yik Yak that stands out most everyone is the ugly one.

Angela, the student with the secret Twitter account, mentioned how her brother started going on Yik Yak at various campuses when he was deciding where to apply to college; he was so stunned by what people said that he crossed certain campuses off his list.

Another student at Angela's evangelical Christian college, Alex, said something similar about his own campus. Alex is kind of an oddball among his peers—skinny and gangly, he makes funny faces while we talk and at one point tells me, in his best pseudo-suave voice, that he's "not smooth with the ladies." At another point, he explains that he is concerned about "where this is heading"—meaning social media in general, but especially anonymous sites like Yik Yak. All of social media, he worries, is "getting more and more like Yik Yak." "It's funny at times, because people post stuff that's pretty hilarious." But the anonymity of it is "kind of scary," Alex thinks. "A lot of people just don't care, you know, what they post about the school and stuff like that. And they're just going crazy on there. They're anonymous though, so it doesn't really matter, they think."

Alex thinks it matters what people say, though. A lot.

"I can see this evolving into something like people posting threats toward other people, and/or giving away private information on there about, you know, someone or something, and it could be used very harmfully and could spread like wildfire, and you don't know who posted it," Alex says. "Once you post something, everybody knows about it. Everybody's on Yik Yak all the time, and so that's a scary thing. That's worse than Facebook or Twitter because, I mean, people watch what they say [on Facebook and Twitter] because it's attached to their face and everything, but now you've got this anonymous thing, so people are like, 'Oh, I can post whatever I want,' you know. It's not cool."

Yik Yak, and the negative exposure it allows, can really hurt a person "because it's seen by everybody on campus," Alex thinks. He feels like universities really need to start paying attention to Yik Yak if they haven't already. "Because it's anonymous, anybody can post anything. They're like, 'Oh man, this power!' you know? 'I just spread terrible ideas without any consequences!'. . . . I see it becoming a bad thing soon."

ANONYMITY: IT DIVIDES US

While anonymity can be problematic, there's no doubt that students crave it.[5] Many of them long for the freedom to be themselves without worrying about future repercussions. Sometimes, they just want to be silly and have fun. Social media is an enormous part of their lives now—it's nearly inescapable—so seeking relief from having to appear happy and perfect all the time not only makes a lot of sense, but it seems healthy.

Students are very much aware of the potential for anonymous sites to foster sexism, racism, and bullying (more on bullying later), and they are extremely concerned about all of this. But they also seem to feel that these are secondary issues. What they are mostly concerned about is that they are locked in to social media as a means of communication but don't feel they can actually be themselves unless they are anonymous. They are so afraid of some future human resources person finding an offending post from five years earlier that they long for an outlet for their true feelings, a place where they can express their emotions without fear of dire consequences. It's no wonder they crave anonymity and freedom from responsibility.

Apps like Yik Yak and Snapchat are offering college students a much-needed playground. They may be imperfect vehicles for this desire, but at the moment it's all students have, and they don't want to give it up.

But in light of this trend, we must ask ourselves what it means that so many young adults feel they can only be honest about their opinions or their less-than-serious sides when there is the promise of anonymity or that their post will disappear within seconds. Students seem to experience the world of social media in terms of extremes: either what you say could haunt you forever (i.e., on Facebook) or what you say is utterly free of consequences (i.e., on Yik Yak). The options before them seem to be curating the highlight reel versus letting one's ugliest thoughts fly for everyone to see. Even the students who regard themselves as "lurkers" on social media are pulled into this unhealthy zigzag as they witness the frenzied display of perfection and joy alongside the equally frenetic display of darkness, bullying, violence, and hate. We must also ask ourselves what it means that these extremes are being exacerbated *during* the college experience, a time when students are supposed to be asking big questions and learning to critically analyze (what are usually false) dichotomies exactly like this one. The humanities and social sciences are supposed to help us develop a sense of nuance and an awareness of the tremendous complexity that the particularities of gender, race, class, ethnicity, sexual orientation, religion, politics, education, privilege, and poverty bring to bear on our opinions, our experiences, and our choices (or lack of them) as we grow up and become citizens of the world in our own right. We need to take the resources we have for thinking about so many particularities and diversity in general and apply them to what students are now living online.

Today there is the physical college campus, and there is the virtual one. In the virtual one, extremes and intolerance seem to reign. And whereas Snapchat is certainly the kinder, gentler (and funnier) option for young adults today, the rise of Yik Yak should concern everyone. Students certainly worry about it, even as they use it and read it. Yik Yak seems to have become a forum for bullies and trolls rather than one for provocative opinions and playful commentary. It's no wonder that college students are so afraid to show any sign of vulnerability to their peers and their larger publics that they hide their realities beyond a veneer of perfection and positivity. Signs of vulnerability not only may sink you with your future employers but also might end up on the anonymous feed of your university's Yik Yak, to be ridiculed and picked apart in front of an audience of your peers.

But how often does that negativity turn into outright bullying? After so much talk of the darkness on Yik Yak and the vicious things that students are saying about each other on the app, the subject of cyberbullying has never seemed so important, and it's the subject that we will turn to now.

7

AN ACCEPTABLE LEVEL
OF MEANNESS

THE BULLIES, THE BULLIED, AND THE PROBLEM OF VULNERABILITY

*I do not like anonymous social media sites and do
not participate in them. I think they are an avenue
for online bullying. People can be very cruel and
heartless on these sites because their name is not
attached to what they are saying.*

> *Tania, sophomore, Catholic university*

*I mean, there's a certain point where, you know, [Yik
Yak] becomes just blood sport.*

> *Ian, junior, private-secular university*

CORBAN: FINDING THE LIKE-MINDED ONLINE

When Corban walks into the interview room, I do a double take. What
in the world is hanging around his neck? I try not to stare. Corban is

blond, short, pale, and skinny, with shifty eyes—he doesn't really look at anything for more than a second or two—and has what look like goggles hanging around his neck. Swim goggles? Or old-fashioned aviator goggles? I can't tell. But I do know that the lenses are bright yellow, and they are big—probably too big for swimming, now that I think about it. He sits down across from me, eyelids fluttering almost constantly, and we begin our conversation. I soon learn that Corban is a first-year electrical engineering major. Corban is also a non-Greek at a school where nearly all the students are involved in Greek life—though he barely seems aware that this makes him stand out. Corban is in the robotics club and the cybersecurity club. As our interview moves forward through the beginning questions, I finally can resist no longer.

"What do you have around your neck?" I ask.

Corban holds the lenses up to his eyes. "Oh, these?"

I watch as Corban blinks at me from behind yellow plastic. "Yes, those."

"These are goggles—I made them," he tells me and proceeds to put them all the way on, pulling the strap so it fits snugly around the back of his head. "I've also made gloves and chain mail."

Chain mail? I'm thinking as I sit here, staring at Corban, who's looking back at me through his goggles. "You make this stuff . . . . for fun?"

"Just for fun," he confirms.

Corban continues to wear the goggles for another fifteen minutes or so before sliding them down around his neck again. During this time he informs me that he is currently on Facebook and Tumblr, but that his first social media account was a "Lego forum." I also learn that his Facebook account was started not by him but by his parents because they wanted to be able to tag him in photos. Corban likes to post photos of his "inventions" online. He is, needless to say, not your average college student. And as he explains these things to me, I find myself worrying that I will soon learn that Corban is a target for cruelty on social media.

It turns out I am—very happily—wrong.

Corban is a very well-adjusted social media user. There are a lot of people out there who share his interest in chain mail and robotics, and social media brings them together. And though Corban may seem like an atypical student in terms of hobbies and appearance, when it comes to social media, he's very much in the mainstream. He has rules about posting ("If you wouldn't say it or do it in person, don't do it online") that are common, and he generally sees social media as a fun way to kill time. When I ask him whether he worries about online bullying, or the ways that people can be mean to each other on social media, he says, "It's definitely a thing I've seen happen. I haven't done it, it hasn't been done to me. And it hasn't been done to my friends personally. It's hard to combat something that you're not directly impacted by or that those around you are impacted by, so it's more of a thing I'm philosophically against but have no way to actually combat."

This, too, is pretty standard. He's never been bullied, his friends haven't either—but, sure, he worries about it in an abstract way.

Corban goes on to tell me that he never really spends time comparing himself to others on social media—which is unusual but healthy. And when I ask Corban if he thinks that social media can affect people's self-esteem, his answer is even more interesting. "I think it can, but really, I think a lot of who you are on social media is a reflection of who you are normally," he says. "So if you're prone to have lower self-esteem normally, on social media things will tend to lower your self-esteem more, whereas if you have a higher self-esteem regularly, social media will help improve that. I generally have pretty good self-esteem, I would say." Then, in a very measured, logical way, he explains that posts that get a positive response help him to feel better about "what he's done," and posts that get a more negative response, he thinks "you can take it with a grain of salt."

Corban gets right to the heart of the problem with online bullying on several levels—so much of it is about keeping perspective, about self-esteem, and, in many ways, whether your skin is like Gore-Tex and the nastiness just rolls right off, or if you're one of those unlucky, sensitive people, and meanness and cruelty go straight to your heart.

THE RARITY OF (NON-ANONYMOUS) BULLYING

Online bullying, or cyberbullying, is a hot topic.[1] Stories of teens committing suicide because peers harassed them so constantly and so viciously terrify and outrage us, especially the parents of children approaching the age when they will set up Facebook and Instagram accounts and launch their vulnerable, tender selves into the wild, uncontrolled, and sometimes extremely cruel world of social media. That world can be soul-crushing for a child or young adult, humiliating in a way that is so severe they cannot see beyond it, so horrific that they cannot find a way to survive it. Online shaming is not restricted to children and young adults, but it is particularly frightening at that age, and many parents and even schools and colleges aren't sure how to respond to it or, ideally, prevent it from happening in the first place.

Just about everyone is aware of the high profile cases of teens committing suicide because of online bullying, tragedies that inspired the writer and journalist Dan Savage (and his partner) to post the first "It Gets Better" video on YouTube in an effort to let other young adults suffering bullying and harassment know that there is hope in the future. Dozens and then hundreds and eventually tens of thousands of similar videos followed, and now the It Gets Better Project is a formal organization that offers assistance to the LGBTQ community and family members of bullying victims. But of course, while there are incredible resources out there for young people, bullying and cyberbullying continue to be terrible and tragic problems, often caught only when it is too late.[2]

The reasons for everyone's concern are obvious. Two separate studies show that cyberbullying seems prominent among both middle school students and young adults—with approximately 30 percent identifying as victims of both cyberbullying and traditional bullying, while approximately 25 percent identified as being perpetrators of both cyberbullying and traditional bullying. Girls are more likely to be the victims of bullying, and boys are more likely to be the perpetrators of it.[3]

And college students are just as upset as parents about online bullying. When I asked the interview participants if they worried about online bullying and the ways people can be cruel to each other on social media, just about everyone said yes. Occasionally someone added that they were especially worried about their younger siblings because kids can be really mean to each other. And students definitely feel that anonymous sites like Yik Yak are expressly designed to promote bullying.

But when I asked these same students whether they've ever known anyone who's experienced online bullying, or were ever bullied themselves, almost everyone would pause, think about it for a moment, and then say no.

"Oh, absolutely," Elise tells me, she is definitely worried about bullying. But then she can't think of anyone she knows who this happened to. "I'm trying to think if I know anyone personally, or like, from friends and stuff, about people being bullied online. Because you know, in high school we'd always have the assemblies about people who are victims of bullying, or even people who were victims of domestic violence via social media. I can't think of anyone personally close to me. I think people are getting smarter about it than they were . . . . but, yeah, I think it's definitely still a big concern."

During my visit to an evangelical Christian college in the Southwest, Max gives me a similar answer when asked whether he worries about cyberbullying. "All the time, yeah," Max says. "I have never experienced it firsthand, and I've never dealt it out firsthand. I *do* think that people feel much more safe behind a screen than they would ever feel in person. So in terms of bullying and in terms of bad comments, they feel completely okay doing it, whether it's under a false profile or if it's on their own profile, and they're commenting and criticizing and bullying, they just feel safer behind the screen because they can't be touched." Max goes on to say that the way in which people feel safe saying cruel and terrible things to each other because they are behind a screen creates "a culture of cowards," which really bothers him.

But he still can't think of anyone he knows who's been a victim of cyberbullying.

Then there's Brandy, who has a somewhat different reaction. "I feel like so many people make cyberbullying to be this huge thing, and I never see it on my personal Facebook," she says. "I never see people doing that stuff on Facebook, and I always find that really strange. People talk about all these cyberbullying cases, and I can't think of an instance where I've ever seen that because I feel like my generation has ingrained in their head that other people can see what I'm doing, kind of thing, which is a scary thought. But I feel since they know that, then obviously you're taught that bullying is wrong, and because of that, people aren't going to do it."

If cyberbullying is happening, she thinks, it's happening privately because everyone else knows that people's online behavior is being watched. If you're worried about future employment (as just about everyone is) and appearances (as just about everyone seems to be), you'd be crazy to behave that way in the online equivalent of broad daylight.

One young man, a football player at a Catholic university, went so far as to say that one of the issues he worries about most is that meanness on social media is so common that we're becoming "desensitized to it." We don't even notice it when it happens, because we've accepted it as a fact of life.

Occasionally, I interviewed a student who thought she had seen people being bullied firsthand but still couldn't name any examples. One young woman, Sarah, was extremely concerned about bullying and raised the topic over and over during her interview. Sarah was on the track team and in ROTC, and she was a resident assistant, too—clearly an overachiever. Sarah applied the Golden Rule when it came to her criteria for posting on social media. "If I wouldn't want [something] posted about me," she said, "I would never post it about someone else." A very self-possessed young woman, she'd never been bullied, didn't worry about ever being bullied, and didn't have any friends who'd been bullied. But Sarah felt as though bullying was occurring all over the place, especially on Yik Yak. She told me she thought Yik Yak was designed as a platform for people to bully others. Overall, she felt, social media "does more harm than it does good."

"There's so much negativity online and so much bullying that it has a negative effect on people's happiness," she said. Sarah went so far as to say that if she saw people being mean to other people she knew online, she "took it personally."

The rise of Yik Yak seems to have drastically changed the landscape—and visibility—of bullying. When I was finishing up my interviews and surveys, Yik Yak was still in its infancy, but even then students had begun to regard it as the "bullying app." And they are deeply upset by it. They worry that Yik Yak is bringing bullying close to home and hurting people they know and love, and they said as much in the online survey. Of the students who chose to answer an open-ended essay question about anonymity, a third of them used the opportunity to comment on bullying, though bullying was not mentioned anywhere in the question. Many students naturally connect anonymity with bullying and opted to vent their dismay.

Now that Yik Yak is raging on campuses everywhere, I suspect that more college students will be able to identify specific victims of bullying—Yik Yak seems to have changed the game. As the app grows in popularity, it's possible that bullying will become less rare and that a growing number of college students will be able to point to many examples.

I did meet a few students who knew someone who'd been bullied in high school or middle school. Amy told me that "online bullying is a big thing." She felt this was because people "don't have to own up to what they say," and the computer acts as a shield—no one ever has to "face the person when they say it." This means they "will just comment random things that they wouldn't actually tell that person in real life." Amy told me that a really good friend of hers from high school had been cyberbullied and that it "made high school a lot harder." Amy was rather scarred by watching her friend go through such hell, and said, "I think [bullies will] do anything to take down their opponent."

The rarest student of all, though, was the one who had actually experienced cyberbullying in her past—and I use the pronoun "her" intentionally. I met only two cyberbullying victims, and both were young women. Though victims are rare, their stories remind us why people are so concerned about this issue. For even one young adult to endure such intense pain, humiliation, and isolation is unacceptable. The two young women whose stories I tell in this chapter struggled mightily to survive and move on from their experiences—indeed, they were still struggling when we met.

MAE: LONGING TO BE DRAMA-FREE

"Cyberbullying, it's very idiotic," Mae tells me about midway through our interview. Normally soft-spoken, she now speaks passionately "I don't believe in bullying, period. I was always bullied as a child, so I'm against bullying completely. I don't understand why people cyberbully anyway because they're just hiding behind a computer screen instead of confronting someone about their feelings or their problems."

Happily, Mae is talking about this with some distance between her life now and her personal experiences with bullying in the past. A tiny, pretty, first-year student at her Catholic university, she is relieved to have put her experiences with bullying behind her and wishes she could help others avoid the terrible things she went through.

Mae explains to me that about six months earlier she had shut down all her social media accounts. She uses Snapchat occasionally to send pictures to friends, but she doesn't put Snapchat in the same category as public platforms like Facebook and Instagram. Mae feels that she has more control on Snapchat, whereas public platforms can be painful, which was too much for her to handle. When I ask Mae to describe herself socially, she says, "I think I can get along with people and I don't stick out of a crowd, I'm just part of it. . . . I don't really like to attract too much attention, so I try to just stay quiet." Blending in, not making yourself noticeable, is a theme to which Mae returns repeatedly. If you stick out, you might become a target, and Mae never wants to be a target again.

There was once a time when Mae felt differently. She *wanted* to be noticed, she *wanted* to be cool, she *wanted* to be liked in ways that made her feel special and different. So she threw herself into social media. All that wanting and hoping made Mae especially vulnerable. She didn't anticipate how, when you seek attention, the kind of attention you get might be bad, and she learned the hard way. But it takes a while for Mae to admit that her desire to be noticed and liked resulted in an episode of cyberbullying that made her swear off social media altogether.

First, Mae talks about her concerns that people aren't always "themselves" on social media. "I feel like because of social media and because of peer pressure and stuff, teenagers especially, they try to be something other than themselves," she says. "I think maybe that's the reason why I stopped going on Facebook. I felt like I was being someone other than myself, and that worried me. So I just stopped going on Facebook."

I ask Mae to tell me more. "Well, high school was really cliquey," she begins. "There was always groups of friends, like how in the movies they portray the popular kids and the jocks and the geeks. I was in band, I was always in the honors classes, I was always a little nerdier, so on social media, I felt like I should try and be like one of the cool kids. But then I realized that's not who I am, I should just be myself. And then I realized how much people try to change themselves on social media, and it irked me I guess, so I just felt like I didn't need social media, really."

Mae's attempts to "be like one of the cool kids" backfired.

"I would post, I guess, selfies," she says. "I would try to make cool statuses about what I'm doing or who I was talking to. I would add strangers whenever they would send me friend requests because their mutual friends would be the cool kids. Or I tried to talk to people that I didn't know on there." Mae mentions—with regret—that she now realizes that it's dangerous to talk to strangers online, but at the time all she cared about was getting attention—any attention. "I really focused on how many 'likes' I got, how many comments I received, how many people told me that my pictures were nice. Then I realized that it affected my self-esteem and my time because I was always on Facebook, I was always worried about who was posting a new status and how many 'likes' they had compared to my status and how many 'likes' that I received and it was just too much, so I stopped."

Mae mentions time and time again how eventually "she stopped" being on Facebook. Almost every one of her answers ends with her saying something like this—how she got off of Facebook or about how unhappy being on Facebook was making her feel. But eventually she expands on what she means by how she used to "talk to strangers" when she was fourteen and fifteen and, with hindsight, how reckless this was.

"Social media before just encouraged me to make friends with random strangers on the Internet that I would never meet in person," Mae explains. "Some of the strangers would be from different countries, so I couldn't even pronounce their names. They [messaged me]. They would talk about how beautiful my pictures were and how old I was, where I lived. They would also add my friends and then they would also message my friends. And then we would all talk about it and how weird it is, but [my friends] wouldn't, some of them wouldn't even delete that person, they would just block him, so that he was still on their friends list but they didn't have to talk to him. I feel like they were trying to 'up' their friends list, like, the number of friends they have on their list to seem cooler about how many friends they have."

This is every parent's nightmare: their little girl (or boy) is talking to potential predators via social media. I ask Mae if she ever told a parent or an adult about what was happening, and she says no. She explains, though, that eventually, "after I thought about it," she began to worry that older men were messaging her. "It was really creepy so I just stopped. It scared me." For Mae, it was enough that she stopped engaging with them—she didn't think she needed to tell anyone. Her mother already had plenty to worry about.

That Mae would enjoy the attention of strangers telling her she's beautiful and that she's a victim of cyberbullying doesn't seem like a coincidence. Mae was so desperate for attention, she would seek it anywhere, and it's this desperation—to fit in, to feel pretty, to feel special, to simply be noticed—that leads young adults into dangerous territory on social media. Not only does it make them potential targets for anonymous criminals, but any whiff of desperation, of neediness, of true vulnerability—which comes from feeling invisible, as if you don't matter—can make someone a target among their peers. This is especially so during the middle school and high school years, but even during college. When you talk to someone like Mae, it starts to seem as if social media is designed to prey on the vulnerable.

Mae tells me how at one point she had to start taking birth control pills, and she gained weight. All the popular girls at school would post bikini pictures over the summer, so she did too, and then she'd sit there on Facebook, comparing her pictures to theirs. She thought they looked better than she did, so she took down all her pictures because of how ugly this made her feel.

This is where our conversation finally turns to Mae's experiences of being bullied. "I tried to fit in online," she tells me again. "But then I just, I would get into cyberbullying fights. A girl would message me and she would get into a fight with me and then I would get into a fight and then there would be this really big fight over both of our Facebooks, and then in messages, and it affected me because I felt like people were bullying me and I felt like I didn't have any friends."

"Did you talk to anyone else about what you were feeling?" I ask her again, wondering whether it occurred to Mae to tell her parents or another adult.

"I talked to my mom about it, and she told me to just stop responding, but I still did [respond] for some reason. And I talked to my friends about it, who all also had a Facebook, so then they would jump in, and that really wouldn't help the situation because then it just caused the fight to get bigger and bigger."

I ask Mae to tell me more about what this "fight" was about.

"It was over a boy," she responds. "It was sophomore year [of high school]. This girl, she accused me of something and then she would message me all the time, she would call me names, she had her friends message me and call me names and say nasty things to me. So then I would just, I would try and fix the problem by trying to talk to them but they wouldn't, they would just continue calling me names, like bitch, and they would use the 'c word,'—I don't like saying the 'c' word. They would call me a whore because apparently, I guess, [they said] I cheated on [the boy], but I never did, I just broke up with him and he was upset about it. This girl was his friend, so then he told her about everything that we went through, he was exaggerating about it, then she just assumed I was this big, nasty monster and then she messaged me on Facebook."

Because both the boy and the girl who were bullying her went to her high school, Mae would have to see them during the day and then endure their taunts and cruel comments online. "But it's funny because I played softball with her, I saw her during school, she lived in the same community as me, but she would never say anything to me in person. She would just look at me and walk away, she would never say anything to me, but on Facebook she can message me anytime of the day and then call me these nasty names." Mae tried to talk to the girl in person about what was happening, "but she just walked away." Then Mae tried to confront her on social media, "since I guess she felt more comfortable talking about my problems with her on there." But the girl only continued to call her "nasty names."

The behavior eventually stopped because the girl finally called Mae one of those nasty names at one of their softball games, and Mae's father heard it happen. "He went over and talked to her, so then after that she stopped, she left me alone completely because my dad had a talk with her." I ask what her father said, but Mae has no idea. "He didn't tell me, and she, I don't talk to her, I was just playing softball and then I just looked over and I saw that he was walking toward her." Mae doesn't care much about what her father said—just that he said it, and afterward, the behavior stopped. That is all that matters.

Plenty of young adults would be able to endure such treatment from their peers—they might blow it off, they might have enough confidence not to let someone else get to them, they might have a thicker skin. But this is the tricky thing about both bullying and cyberbullying: what one person might be able to endure is another person's nightmare. Some of us are simply more vulnerable than others to such treatment. And in situations where someone like Mae is preyed upon and doesn't have the emotional resources to withstand the experience, that the bullying would be amplified, expanded, and made even more public on social media makes it that much more destructive. The bullying can seem so big and so pervasive and so impossible to escape—it's not as though it happens only when that person is in the room, or stops when you leave the school building—that someone like Mae starts to feel like she's drowning.

This means Mae is not a good candidate for social media, and she knows this now. Granted, she thinks that social media *can* be good—for socializing with family and friends who are far away, for example—but the drawbacks for her far outweigh the potential benefits. She is unbelievably relieved not to be on it anymore and not to be subjected to the potentially cruel treatment of her peers as a result.

"I don't have to deal with the drama," she tells me. "Because I feel like people express their drama on social media rather than talk about it in person, and I feel like if I [go onto a] social media account, that's all I would read, just drama. And I don't want to read drama, I want to read the good things that people do during the day or achievements or good news," she emphasizes. "I feel like I'm drama-free [now] and I'm not wasting my time sitting at a computer screen, messaging people or posting statuses or, hashtag this, hashtag that."

I wonder whether she ever misses it or if it's difficult to be completely off of social media now that she's at college and everyone around her is on it—and using it to socialize—nearly constantly.

"No, I honestly don't," she says. "Now whenever I listen to my parents talk about Facebook or I watch people get on Facebook or Twitter or whatever, I think about what I used to be like when I had a Facebook and then I realize that I enjoy not having a Facebook now, more than I did when I had a Facebook. And then I look at them and I realize how much time they're wasting just sitting there on social media when they could be going outside, and hiking, or spending time with family, studying for school." For Mae, coming to college has actually been a relief because so much of the socializing happens face to face. This is much more suited to who she is. "I feel like [college] has helped me become less socially awkward," she says.

It's also true that just about everyone she's met at college is constantly on Twitter and Instagram, they're always updating and posting pictures, and sometimes Mae feels a little pang that she's disconnected from something that consumes everyone else around her so completely—but then she realizes that she really doesn't want to go back there. "Sometimes I feel like I'm missing out, when everyone's talking about that funny tweet or all this other stuff, but then at the same time I *don't*. I *don't* feel like I'm missing out and I feel like I made a good decision to stop social media, to stop using it." I ask Mae if she thinks she might eventually change her mind and go back onto Facebook. "I think I can make a strong commitment to [not doing this]," she answers. "It's been a while since I've gotten on Facebook, and with how busy I am in school I feel like it distracts me from using social media. But whenever I do get bored and I have the urge to just browse on the computer, I just go and find someone to hang out with, I just find something to do." It helps, too, that the friends she's made at college have accepted the fact that Mae just doesn't do social media.

The cost of social media was too great for Mae, and the benefits of quitting have been too big to ignore. She feels like a much healthier person now. "I feel like how I see myself now has positively increased. I don't care what people say anymore, I focus on myself, I don't have low self-esteem, I'm comfortable with my body, I'm comfortable with how I look and who I am, so I feel like it has helped me a lot."

HAILEY: FACEBOOK IS NOT A PLACE
FOR VULNERABLE PEOPLE

When Hailey walks into the interview room, I immediately think of Sarah Jessica Parker's character on *Square Pegs*, a television show I used to watch when I was a kid. Hailey has long, curly black hair, glasses, and the kind of geeky, funky style that Parker embodied as the lovable Patty Greene. Hailey is incredibly sweet and also incredibly insecure—despite the fact that she is a physics major at one of the top private universities in the United States. Sometimes Hailey is so insecure that it's painful to listen to her talk.

Right out of the gate, her face falls and she tells me that college has been *really* rough. Then she brightens and talks about her major—particle physics. She "likes studying neutrinos," and she has met pretty much all her friends through her major. What makes her happy is having deep conversations with them about ideas. And Hailey finds meaning in her work. "I *really* like doing physics, doing my research," she says. "I know that I'm helping expand research and expand our knowledge in this world, and hopefully later on down the line, that will help people in a direct way." It's obvious that Hailey loves her studies, and at least in the academic sphere, she has some confidence.

But as soon as the topic shifts again, her confidence evaporates.

The reason college has been so rough is because of her boyfriend—now her ex-boyfriend—who was abusive. He made her convert to Christianity, but she was never good enough in his eyes, she tells me. The boyfriend did not like that she studied physics—the one thing Hailey truly loves. "He thought that God wasn't calling me toward it, and he was trying to use that to try to make me doubt that I should do physics."

Hailey lets out a big sigh, then apologizes for "dumping" her boyfriend troubles on me. I tell her not to worry, and we begin discussing her online life. At the moment she's on Facebook, Snapchat, and Yik Yak. Like so many other students, Hailey talks about the importance of posting only happy, funny things. "I want to show people that I'm happy and that my life is good," she says, even when it isn't. Hailey posts happy things even when she's unhappy; if she's too unhappy to show the world she's happy, she doesn't post at all. Besides, her boyfriend used to tell her it was ridiculous when she posted silly things or tried to be silly in public. It only made her look stupid. "Nobody wants to see you making a joke," he would say.

By the time Hailey finishes explaining this to me, she sounds morose. I ask if she has stayed away from social media since she's been having such a rough time, and she responds, "I unfriended the ex-boyfriend, and, yeah, definitely I stayed off of Facebook a lot." Hailey explains that Facebook was one of the reasons they broke up. He refused to make their relationship "Facebook official," plus, others girls were flirting with him on his page, which hurt Hailey's feelings. "I told him when I unfriended him that it was because these girls were flirting on his Facebook, that I just didn't want to see it. . . . I just don't want to see it." Then, there was the fact that her boyfriend hated when she tried to be funny on Facebook—she repeats this four times during the interview—it embarrassed him, which made her feel ashamed.

I ask Hailey if she's ever considered getting rid of Facebook altogether, and she tells me, yes, actually; very few of her physics major friends are on it, so sometimes she doesn't see a point in staying. And even though part of the reason she's stayed is to keep in touch with people from high school, some of those people have hurt her in the past. "Actually, in high school, it was really, there was a lot of, you know, you see things on Facebook and it upsets you, or somebody makes a status about you that's mean."

Hailey was bullied in high school, she explains. Some girls would post videos on her wall to make fun of her, and the girls at her school would call her a lesbian, even though she isn't one. "They would make jokes saying, 'Oh, you're like, what is it like to be a lesbian?' and blah blah blah. I didn't identify as a lesbian at all, but also that was mean of them to say. . . . That's insensitive." Hailey had another friend who these same girls would encourage to "kill himself." Recently, she went back through her Facebook timeline from high school. She feels a lot of shame about it—not only about having been bullied but also about the way she responded. "A couple months ago I looked at the stuff I had posted when I was a freshman in high school and I was like, 'Oh my gosh. I'm so embarrassed.' I posted stuff about people bullying me online, and like, I would not, I mean, I guess, I don't know. I posted stuff and I was like, 'Oh all the bullies. I'm talking to the principal tomorrow, and all the bullies are going to get their just deserts,' or something like that."

She sighs. "Nobody responded or commented."

Hailey thinks that social media is not a good place for "weird people" to be their weird selves—and she is weird, she says, or at least she was in high school. She believes this was why she became a target. Weird people are particularly vulnerable, in her opinion. It's too easy for others to go on their Facebook pages and make fun of them. "That's also why I don't like being vulnerable on Facebook. Other people can look at it and say, make fun of me you know, because you're not there. They're seeing this . . . and, I don't know, they can be anywhere. Anybody can see it."

Basically, Facebook can become a world of hurt for certain people, according to Hailey, because of all the ways in which you discover you've been left out of things, because of all the drama that can come from it, because of the ways people are mean to you. "I went to a really small high school, like, thirty-five people in my class. Toward the end of junior year, everybody was sort of friends with each other, and then all this drama started happening. Basically, there was a faction, and I tried not to choose a side because I saw both sides of the story, and I was like, 'You guys are being ridiculous. Each side is misunderstanding what's happening.' I guess, because of that, a lot of the girls didn't like that I was doing that, and so there was a hierarchy formed, and there would be parties. Like, they would invite my two best friends, but I wouldn't be invited, and I'd see the pictures on Facebook. And that was pretty hurtful. It was sort of a way to make it clear, 'Oh, you're our friend but not that good of a friend.'"

Later on, Hailey circles back to the subject of how Facebook can make you feel bad about yourself. "I feel like Facebook can really hurt me and make me feel lonely, if I see things on Facebook, like, 'Oh, these people are having a really good time.' If I'm in a bad mood and I see people having a good time, that can make me feel really lonely. . . . I think it helps to have friends who are off Facebook, and who still communicate with me, even though they're off Facebook because it's nice to know my friends aren't on Facebook, they're in real life. They're *here*. . . . But then, yes, there's still a loneliness factor for me, and I think it probably is different for different people." Hailey cites her current roommate as an example of someone who seems made to succeed on Facebook, for whom events haven't really happened until she posts about them. Her roommate loves Facebook and is constantly updating her page. It's people like this who gave Hailey a hard time about not making her relationship status public, which made Hailey feel terrible. "It's like, if it's not on Facebook, it didn't happen, or it's not real," she adds.

Leaving people out of parties has been going on as long has high school has existed. The difference today is that young adults end up living out their humiliations in public. It is thrown in their faces—sometimes unintentionally, but sometimes *very* intentionally. Social media becomes a tool for provoking hurt among the more vulnerable, or the "weird," as Hailey put it. Savvier young adults learn to use it as a kind of weapon, and for girls like Hailey and Mae, who want badly to fit in but don't quite know how, social media can take what once might have been a private, fairly brief episode and turns it into something that lasts forever, continuing to victimize them far into the future. That Hailey could go back through her Facebook Timeline and see evidence of this was shameful and upsetting.

It's also true that many students feel pangs of hurt when they see pictures of their friends hanging out without them, or pictures of people at their college partying and having a good time, when they feel like they never get to—maybe because it's not their style or, more likely, because no one told them about the party. Again, this is pretty typical social politics for people of that age. But students feel many different levels of hurt about this, and for someone like Hailey, the hurt is deep and lasting, especially because of how public everything is. Social media announces to everybody around you who is included and who isn't. Each time you go online, there is evidence that you aren't cool, that you aren't desirable, that people don't even know you exist. People like Hailey aren't as emotionally equipped as others to handle this constant barrage. Hailey has a thin skin, and social media requires young adults to develop thick skins—very thick—at very young ages, so that they can withstand the disappointment and hurt that can come from being exposed to the nasty side of this "connecting."

Despite all of this, Hailey made it through the dark days of being bullied in high school; she still maintains her accounts and says that she likes the ways in which social media connects her to people. But it's obvious when talking to her that at one point in her life, being on social media was devastating. The way her voice wavers, the way she backtracks again and again to mention her shame and embarrassment is painful for me to witness. One would hope that Hailey, who is incredibly smart, could find a way to better interpret her experiences on social media. The problem, of course, is that social media tends not to be an intellectual enterprise but an emotional one, and on that level, Hailey simply falls apart.

JACK: THE VARIETIES OF TROLLING

As rare as it was to find anyone who had been bullied, it was just as rare to find anyone who would admit to being mean to others online. Of course, it's not surprising that it would be difficult to find people who freely admit they're awful to others. It's unbecoming to identify as a jerk, even when you're promised total anonymity. For someone to sit there, across from an interviewer, and say, "Actually, *I'm* one of the people who's nasty to everyone else," is a pretty intense thing to do.

But I did meet a few people who owned up to this behavior—all of them young men. One of them, Jack—a supersmart, academically engaged senior at a northeastern private-secular university—seems like an incredibly nice guy, is thoughtful in all his answers, and always responds with a careful critical analysis. He's very interesting to talk to, so when he tells me he's an Internet troll—someone who posts provocative comments to get a reaction out of others—I nearly fall off my chair.

Jack's training as a troll started early—back in middle school, on Myspace. "I was trying to be a sort of edgy comedian, and at that time, you now, edgy comedy is sort of this very . . . . Jack pauses to sigh. "It's just, you write all the things that you think would offend somebody, and you just cram it all together and write basically the most disgusting things you can think of." He pauses again. "And those are things that you shouldn't be putting your name on, right? . . . . And, you know, I really didn't like high school and middle school, so I definitely aimed a lot of this, I guess you could say *aggression*, toward the established school and that sort of thing. Not like death threats or anything crazy like that," Jack adds.

On one level, Jack sounds a lot like any kid who feels alienated and frustrated, even angry, by the politics of middle and high school, like someone who feels left out and knows that he doesn't fit in. But whereas the attempts to fit in by young women like Mae and Hailey turned them into targets, Jack lashed out instead.

Jack's comments on Myspace were aggressive enough that his peers complained to their teachers and the principal, and his parents freaked out when they saw what he was posting. Jack thinks people at school must have read what he said and thought, "He seems to be a little bit too mad at school. And because, you know, you've got all of these creepy people who say these terrible things about school and then *do* terrible things," everyone became concerned. "In reality," Jack goes on, "I was trying to do this sort of satirical thing, which *I* thought was satirical, but someone reading that would not probably think that." As a result of Jack's posts, the school suspended him, and his father installed a program on the computer that monitored Jack's every keystroke.

Now that Jack's older, he knows better than to sign his name to any-thing inflammatory, though he tells me that, as an atheist, he enjoys pick-ing fights with religious people on Facebook—elaborate fights that go on and on and that people get upset about. "Sometimes," he explains, "I'll read a post, and I will think it is so outrageous that I need to say some-thing, and then I write out a whole argument." The behavior is nearly compulsive. If someone says something "stupid," he feels it is necessary to tell them they are "incorrect." Sometimes, after Jack writes out a very long comment in response to something he thinks is ridiculous, just as he's about to press "enter," he thinks better of it, but at other times he can't resist. He goes on to give me a very long example of the kind of fight he picks. People get *really* upset when Jack posts this sort of thing, and it's dismaying to him that so many people can't handle being con-fronted about their beliefs and opinions. "So I think [with] things like online trolling," he says, "you can get mistaken for a troll, even if you're just trying to do constructive discourse."

When I ask Jack to elaborate on getting "mistaken for a troll," I learn that, in his view, some trolls are good. Jack thinks he is giving social media something it desperately needs, something virtuous. "In some ways, de-pending on the troll, it could be really good, meaning that I think, to a certain extent, I can be a troll in conversation," he says. "Now, now don't mistake what I'm saying. Now there's the *trolls*, right? The really, just evil, sick, sadistic, online bigots, and that's not what I'm talking about. I'm talking about the contrarians, right? The people who just argue with you because they can and because they want to find something wrong with what you're saying, even though they don't see it. So a troll could just be somebody who's a devil's advocate, right? And I think that's what a *good* troll is. A good troll is someone who is going to disagree with you for the sake of disagreeing with you, and then just trying to find the flaws of what you're saying. So I think that you can take something good from troll-ing, right? Which is something you probably don't hear very often." Jack pauses, then backtracks once again. "But the people who just blindly hate or blindly dismiss arguments for no reason, I mean, that's what more tra-ditional trolling would be in the public mindset these days. I don't think trolling was always that. I think that trolling has sort of developed over the past, I guess, fifteen years now, where, you know, being a troll could just be posting a funny picture or being a troll could just be purposely evoking a reaction from somebody, but trolling could also be publicly shaming or hating somebody, and that's not how I always thought about trolling, but nowadays that's where [trolling is] going."

Jack is right: I don't hear many people talking about "good trolls" or any other kinds of trolls for that matter—in fact, Jack is the only one who espouses such a theory. I'm impressed that he is so honest about it, and as he parses out the types of trolls, from best to worst, Jack's personal stake in this theory begins to come through. Trolls are *not* just people who say negative things, Jack emphasizes. Trolls, in many ways, are *misunderstood*, and by the end of our interview, it is obvious that Jack feels he is a misunderstood troll. He's just trying to be himself—which is to be argumentative, and to take people to task—and most people, unfortunately, do not find social media to be a good forum for such behavior. They get upset when someone disagrees with them.

The irony of Jack's self-understanding, of course, is this: Jack definitely knows that over the years, many people have felt bullied by him, considering him inappropriate, hurtful, and insensitive. But Jack is also a smart guy who likes to argue, and he doesn't quite understand why he can't just "be himself" on social media, why there isn't space for someone like him, and why there is little to no tolerance for the kind of provocative commentary he enjoys so much. And though Jack clearly doesn't want to be seen as the bad guy, he now knows that to many people, that is exactly who he is. His defense of the different types of trolls is really a defense of himself and his broader identity. The failure of social media to tolerate someone like Jack, to make space for how he's different from most people (who are all trying to be positive and happy), is incredibly alienating for him. One could argue that the ways people on social media have tried to shut Jack down and close him out is a kind of bullying experience of its own. There isn't room for someone like Jack, and he knows it. Social media isn't kind to people who do not conform.

Jack has changed his ways online. In our interview he talks at length about how he's learned to keep his real, provocative self hidden, and is careful now about anything attached to his real name. "A sort of whitewashed PC version of me exists online, very purposefully," he says. "When I was in high school, I didn't understand that you should only put PC whitewashed versions of yourself online." The sarcasm and anger he feels about this are evident in his voice. Jack's social media accounts are now "more of a tool than an identity, I think . . . it's a very conscious crafting of what, when you type in my name, I'm going to very, very purposefully put what I want you to see up there."

He might finally be toeing the line. But he's not happy about it.

IAN: TAKING PLEASURE FROM OTHERS' PAIN

When Ian struts—and I mean struts—into the room, I realize that our conversation is probably going to be a little different than the ones I've had so far at his prestigious university. The vast majority of students I've met are serious, studious, dedicated, thoughtful, and rather sweet, the kind of young woman or man you'd imagine might attend a top-notch academic institution such as this. But Ian has swagger, and he also has the look of a stereotypical frat boy, complete with a bit of a beer gut sticking out from underneath his T-shirt. Sure, his majors (physics and applied math) are as demanding as those of his peers. But while his fellow students are thoughtful, reflective, occasionally socially awkward individuals, Ian seems like a guy who will say just about anything, including something potentially offensive, and then laugh it off like it's no big deal.

In other words, Ian seems to lack a certain level of self-awareness or, at the very least, isn't self-conscious about how he comes across to other people. He's the same Ian who opined so colorfully about women's propensity to take far too many selfies and share far too often.

Soon after we start talking, I learn that Ian actually is a frat boy—he's a junior now, but he started pledging a frat a mere two weeks into his first year of college. He absolutely loves being in a fraternity and speaks at length about all the amazing people he's met because of it. It's also clear that Ian is just as serious about his studies—he's thrived academically and loves being surrounded by smart people. Then Ian tells me that he's an Eagle Scout and is still heavily involved in the Boy Scouts. "I work with a Boy Scout troop," he says. "I'm very involved. I might become an assistant scoutmaster, but I have to go through some training. It's a little bit of basic first aid, but also certain things like boundaries. So they like call it risk zones, so you know what to recognize if a kid, you know, is being bullied or something."

So Ian is getting trained in how to respond to bullying—it's one of the first topics he raises.

Ian goes on to talk about his hopes for the future—to be a good father, to have "a house big enough that can handle [his] hobbies," to have a "project car" in the garage, a workshop there, too, and a place for his model trains. Ian loves model trains and lights up while talking about them. As Ian continues to discuss these things, he loses that swagger and frat boy–type attitude, seeming more like a sweet, excited kid who loves his majors and is thrilled by studying. But the swagger returns when the topic shifts to social media.

Ian rolls his eyes about all the embarrassing things that people put up on Facebook when they're in middle school and high school—and he's no exception. "Any sort of true candidness that people can pull out on you can be embarrassing, you know?" Ian is referring to the kind of self-revealing posts that younger kids put up, especially ones that reveal their enthusiasm or lack of social savvy. I press Ian about why he thinks people shouldn't be candid. "We all know people online who embarrass them-selves," he explains. "And half the reason we all go on Facebook, I think, is for the glimmer of hope to see, like, somebody's engaged now, or, you know, they go on these political rants or something, and it's just some finger-licking good gossip that you get, you know? My friend is from Arizona, and he comes from a little, little town, and he shows us the silly things that some of the people from his very, very small town—you know, not many people ever get out of the town—post on Facebook. Some of them are hilariously open, like taking a picture of their parole ticket and posting it on Facebook . . . or being friends with their probation officer, you know, saying they're going to go smoke some weed or something. . . . I think there's obviously a certain filter that you want to put up."

That people should have a "certain filter" online sounds reasonable, and I believe most people would be at least a little shocked by some of the things Ian describes. But the it's the *way* Ian discusses these things that is remarkable. He is positively gleeful. He knows he is superior to these people, and he loves the feeling. As Ian mentions how "half the reason we all go on Facebook" is to find that "finger-licking good gossip," he gives off the impression that he just can't wait for someone to do some-thing stupid; the moment they do, he's there, whooping and cheering and laughing at someone else's shame.

As our interview continues, Ian's harsh attitude becomes even more evident. I press him on whether he thinks the primary reason people go on Facebook is to look for gossip, and he tries to backtrack, but then he completely undermines his effort.

"I don't think it's necessarily that *I'm* trying to make fun of people or anything like that," he says at first. "But there is absolutely an element where it's like, you know, if there was a girl my age who's just had a baby or something, and she's posting about how excited she is, all of *us* are clearly like, 'This is so clearly not what she was looking to do . . . .and I'm like, 'That's not true. You're lying through your teeth about [being excited].' And I don't know what the adjective would be, but in a sort of primal way, it's kind of, you know, it's *exciting*. It's kind of, like, 'Ooooh!' you know? A little bit of theater going on before me." Ian pauses a moment and points out that he is capable of enjoying the positive stuff, too. "Then, you know, you got other uplifting [posts] where I see someone who just became a firefighter that I knew who struggled a lot in high school and stuff." Ian pauses again. "I wouldn't say I'm malicious in what I look for in Facebook or something, but you know, it's almost like television, where sometimes you watch your favorite sports team and hope they do well, and sometimes, you know, you turn on a reality show and just hope to see, you know, something silly happen, or something that is a little surreal to you."

I ask Ian about the connection he's made between social media and reality TV—he's the first and only person I interview who makes it. "As for Facebook, yeah. If there's somebody in your hometown who, like, got hit by a car in town or something, it's kind of like little mindless news that won't affect me in any way, but it's just kind of silly things that I can keep up on. So, you know, someone who didn't get really hurt or anything but, you know, was on a skateboard and got clocked by a car a few weeks ago or something. . . . It's just the sort of updates that you don't really need, but you kind of want." I sit there, stunned at the callousness he is displaying.

Ian goes on to elaborate about the voyeuristic aspect of social media, telling me that the best voyeurism is found on anonymous sites. "Some of the most juicy kind of carnal satisfaction I get is from the Facebook pages of [my college's] Secrets [and another of my college's anonymous pages]," Ian says. "Every time something kind of scathing comes out, or somebody posts a secret about this or that, even if it starts anonymously, you can see people get into these ridiculous arguments, and, if you're reading through a hundred comments, [it's] either because you know this person in real life or you've noticed trends in their thoughts from the rest of their postings, and you kind of pick ones to root for in the comments if they're all going at each other with torches and pitchforks and stuff. So you'll be like, 'Oh, this girl, she's so annoying. Let's see if they, you know, draw and quarter her here,' and be like, 'Ah, she got them on that comment, but let's see if the next one—Oooh! Oooh!—you know, they got her right back,' and it's great." "I mean," he adds, "I think people are confusing open discourse with just rudeness," even though his pleasure in all this rudeness is obvious.

On the subject of Yik Yak, Ian believes that "it sometimes can pander to the lowest common denominator." "Sometimes you'll see some good stuff, and then sometimes you'll see some just unenlightened shit that's just annoying," he says. "I think it's especially nice to be on the outside of any of the drama. When [Ian names a group of people on campus] are being talked about on Yik Yak, man, I need a cigarette after reading some of those Yik Yaks about them! They were so great! They were hilarious, they were mean, they were nasty! And, you know, everything you want to see in a completely anonymous feed."

The more Ian talks about this subject, the more frenzied his speech and the more excited his facial expressions become, his eyes bright with the thrill of it all. I ask Ian if he ever contributes to these anonymous feeds himself. "Yeah, never with anything totally honest, never with anything really inflammatory or anything," he says. "Because I know that you can really hurt people with that." Although Ian says he would never post anything so hurtful himself, he loves it when someone else does. "It's absolutely fun to witness," he says. "I mean, there's a certain point where it becomes kind of just like blood sport, and it's like, 'Oooh, this is a little too much right now!' you know, and it certainly crosses a line. . . . But, I mean, at a certain point, you've got to invoke the mercy rule on some issues." When I ask if he ever cries mercy on behalf of others, he says no. Yik Yak takes care of that on its own because eventually everything on it gets pushed so far down on the feed that no one sees it anymore, and Ian feels that is mercy enough.

One might argue: Aren't we all a little like Ian? Is it unfair to judge him? Isn't at least some of the appeal of social media seeing the ways in which people make fools of themselves or reveal details or aspects of their personality (like arrogance, smugness, cluelessness) that they don't realize they're revealing? Whether we're college students or not, most people on social media can cite at least somebody they know who inspires the train-wreck effect—what they say is so over the top, so conceited, so lacking in self-awareness that it's difficult to turn away.

But there *is* a difference, I think.

Ian stood out among the students I interviewed as someone who was particularly cruel in the way he took pleasure in the humiliation of others. He could barely contain his glee as he gave me example after example of other people's missteps. Unlike Jack, who may indeed be an Internet troll, Ian does not have the self-awareness and self-critical reflection to realize that trolling might be harmful. Ian is self-righteous about his joy in other people's pain, and more or less assumes that his attitude is the norm.

Perhaps it is. Perhaps the majority of the students I interviewed were just better at hiding it. But I doubt it. The majority were young women and men who—while I'm sure may occasionally enjoy some good gossip via social media—had enough experience with life and with personal struggles to feel empathy for others, or at least sympathy for people who had experienced missteps or difficulties on social media. And though studies like that of Sarah Konrath, Edward O'Brien, and Courtney Hsing at the University of Michigan have famously shown that technology is making us—and especially the younger generations—less empathetic (to the tune of 40 percent), I did not sense a lack of empathy or emotional sensitivity in the students I interviewed.[4] They may all be presenting themselves as perfect online, which may not seem so empathetic, but in person these young women and men are as open and emotional as ever, and they are suffering under the weight of so much pretending and hiding.

Ian, however, didn't seem to have an empathetic bone in his body. This, from a young man with a brilliant mathematical mind, who has aspirations to be a good father, who loves his model trains, is an Eagle Scout, and, ironically, is in training to recognize and respond to the signs of bullying in young people.

SHOULD ANYBODY NEED A SKIN THIS THICK?

True cyberbullying seems a rare thing.[5] It's difficult to find young adults who have experienced it. Rare or not, there is a reason we, as a society, should be concerned about cyberbullying. When we hear extreme stories of teens who have been harassed so cruelly and so constantly they've thought about killing themselves, who cannot see a way through their humiliations, it shouldn't matter whether these cases are rare. That *any* teen or young adult should have to endure such terror, humiliation, cruelty, and total rejection by their peers is unacceptable. We should not tolerate it. Period.

One student, Nora, tells me that she and her sister were bullied online during middle school, and now—from the safe distance of college—she worries a lot about other kids. "It was just negative, hateful comments about how I looked, how I dressed, boys, that kind of stuff," she says. She can talk about it now with an obvious sense of detachment. But the experience makes her angry about the existence of cyberbullying. "When I hear of someone doing it to other people, the first thing I tell people is don't, *don't* feed into it, you need to report it, you *have* to report it." Nora feels that she and her sister became targets for bullying because they posted if they were feeling sad or down. They learned the hard way that "it's better to be positive than to be negative" on social media. Any show of vulnerability opens you up to bullies who will prey on you.

After talking to so many student about this subject, it becomes evident that bullying, in many ways, is in the eye of the beholder. Some young adults are simply more emotionally equipped to brush it aside when people are nasty to them. What might feel like bullying to one person is just something to be "blown off" to another. Students seem to agree that what qualifies as actual bullying, depends, in part, on how the target takes things. Naturally sensitive people and people who show any level of vulnerability or emotion on social media—whether sadness or unbridled enthusiasm—are at much greater risk of bullying, harassment, and mean treatment from others. Just about everyone agrees that younger kids (middle schoolers, for instance) are far more vulnerable to bullying than older ones, and that part of what it means to grow up on social media is to develop a skin thick enough to endure life online. The near-universal mantra that you must appear happy on social media starts to make more sense when you recognize how vulnerability turns you into a target. The appearance of constant happiness is a defense mechanism, a way to protect yourself from the risks that come with putting yourself out there for the scrutiny of others.

Yet even college students who learned long ago to protect themselves on social media are still occasionally surprised when a post sparks meanness and cruelty. Everybody knows that whatever you post is subject to public scrutiny, but sometimes that scrutiny shows up in unexpected ways. And usually this happens on an anonymous forum such as Yik Yak or Reddit.

Like many of her peers, Maria believes that the level of cruelty one might experience "depends on the degree of anonymity associated with the social platform," as on Reddit and another platform called 4chan. She warns me to not even visit 4chan. "It's *very scary*," she says. "It's an evil place because there's so much anonymity . . . .whereas on Facebook, your name is associated with what you say, so you are what you say, and what you post is somewhat representative of you, supposedly. But on these other websites, it doesn't matter what you say so you can say whatever you want." Maria goes onto explain that sites like Reddit have a ranking system that allows people to see what is most popular at the moment. People "upvote" and "downvote" everything that is posted on the site. Huge numbers of upvotes will push a post all the way up to either the general home page or the subject home page. If people don't pay attention to a post, it gets pushed so far down that it falls off the radar completely. Often what's most popular is something very mean, Maria says, and this is "disheartening."

Maria once had a very "successful" post on Reddit—well, successful at first. Then it all went downhill. It was a photo of her in a Halloween costume. "[The costume] was something I made by hand, so I was proud of it, and I wasn't really into Reddit at the time," she says. "I was just kind of using it to figure out what was new in the world of the Internet, and [the photo] got very popular, and I thought there was nothing wrong with it. You know, it's just a picture, and I had thousands of upvotes, but it shows you the ratio of upvotes to downvotes and then your net score. So my net score was, I don't even remember. But the number of upvotes was almost equal to the number of downvotes." Theoretically, downvotes are a way of pushing posts to a place where they no longer matter, of saying,

"I don't want more people to see it," Maria explains. "And then an upvote is like, 'I want more people to see this, I want the world of Reddit to see this.' I think that just comes with having a lot of exposure and gaining popularity. You're more exposed on that platform."

As Maria's post got more and more votes, things started to get nasty and rather ridiculous. "[The photo is of] me in a wig, and my face is covered [she describes her Halloween costume]," she explains. "To me, there was nothing you could say negative about it. But people were saying things like, 'You're fat.' Or, 'Your mirror is so dirty.'" Maria had taken the photo in her parents' bathroom, where they have both a toilet and a bidet. "People were like, 'So rich! Why are you so rich? You have two toilets!' I was like, 'What?!' They were picking apart my image, looking for things to downvote for. It was so weird," she adds, shaking her head and laughing.

Maria is trying to shrug off the experience, to prove she realizes this is the sort of thing that happens all the time when you post things on the Internet. If you choose to expose yourself in any way, however innocent, you must be prepared to suffer the consequences. Maria laughs a lot and rolls her eyes as she recounts this story, but it's clear she was hurt by it— especially the comments about her being "fat"—and she's trying hard to seem like she wasn't bothered by it. Maria tells me she's learned her lesson about posting on Reddit and won't do it again anytime soon because even the most innocuous pictures aren't safe. "I posted a picture of my puppy when I first got her . . . .and people downvoted it! It's just a picture of a puppy! Who downvotes a picture of a puppy?!" Maria's eyes are wide, and she continues laughing. "People!" she adds, shaking her head.

Overall, young adults have come to expect—and accept—a certain level of meanness on social media. They endure it and try to shrug off. Getting mocked, getting harassed, having people say cruel things to you occasionally—by both strangers and "friends"—is just part of the deal. One young woman spoke of getting called a slut again and again on Twitter. While some people might feel bullied by this, to her, this sort of negativity is simply a consequence of being on social media. She didn't feel bullied. The experience seemed to roll right off her.

Bullying is in the eye of the beholder, it seems.

People with thick skins fare better on social media than people without them. But how thick do we want our children's skins to get? How healthy is it to feel you need to protect yourself all of the time? What does this do to a young adult's sense of self? Or his or her ideas about others? Should *anybody* ever need a skin this thick?

Again, college students seem to be living between extremes. Most of them are doing everything in their power to appear happy and display only positive emotions, which—in addition to protecting one's image and future professional interests—is a way to avoid displaying one's vulnerabilities. And there is the minority, made up of people like Ian and Jack, who like to poke at any exposure they can find and, at least in Ian's case, take great pleasure in the despair and humiliation of others. With people like Ian and Jack around, why wouldn't their peers work hard at trying to become invulnerable, or at least present themselves as such online?

What I find odd among the bullies is that they seem to believe that they *are* invulnerable. They don't seem to realize their behavior might get them into trouble at some point, if they push too far. Students are either acutely aware they are vulnerable and therefore constantly work to present an impenetrably happy front as a way of protecting themselves, or they don't have any sense that they *could* become vulnerable by posting inflammatory things because they truly believe they are one of the few invincible people out there.

Conversation on social media—either associated with one's name or anonymous—lacks nuance. It tends toward the very, nonthreateningly positive or the very, sometimes threateningly negative. Emotions on social media lack nuance, too. And while I do not believe the students I interviewed and surveyed lack critical or emotional nuance in their real lives, they are learning to practice extremes online. We should be worried about how this practice will affect their (and our) ability to stay nuanced about these things in the long run.

As my interview with Nora comes to a close, she has one more thing to add. She sighs heavily and says, "If I could just hug all the kids in the world that are being bullied I would."

8

SO YOU WANNA MAKE THAT FACEBOOK OFFICIAL?

I wanted the world to know that I was so happy to be with her. I guess, like, standing on top of a mountain and shouting it out to the world. That was my way of doing that, and I also wanted every other guy out there to know.

Mark, senior, public university

We definitely went through every single social media post, and we're like, "Delete that. Delete that. Delete that," because we wanted to have this image that we were together.

Hannah, sophomore, public university

ADAM: OH, TO BE YOUNG AND IN LOVE (AND ON FACEBOOK)!

Remember Adam from the chapter on selfies? The young man who used to look in the mirror and not like what he saw, but who—after finding a girlfriend—changed his whole outlook? Perhaps not surprisingly, Adam was also effusive about making his relationship public.

If you go to Adam's Facebook page, he tells me, the first impression you'll get is, "Number one, I love my girlfriend because she's in my profile pic." You'll see other things, too, like Adam's favorite city, his favorite football team, and that he likes architecture and has many intellectual interests. But, mainly, you're going to see pictures of Adam and his girlfriend. He jokes that on his page, it's "Myself, her, myself, her," and then maybe a photo of a dog. Oh, wait, the photo of the dog is Adam, his girlfriend, *and* the dog, he adds with a laugh.

Adam and his girlfriend are "Facebook official," of course, meaning that his status is "in a relationship."

"I actually asked her," he says, seeming a bit embarrassed. "I'm not really sure why. I think I just wanted it because I kind of wanted it to be there to remind myself that someone *does* love you. Someone *does* care about you. And then it was also for show because I'm that guy who's never had a girlfriend, so everyone was waiting for it, and as soon as everyone saw it, they threw a party in a sense."

Adam's experience of being Facebook official is incredibly sweet and endearing, and it has done wonders for his self-esteem. Showing off his relationship on social media reminds him he is loved and cared for, in a way that no one before has cared for him, at least romantically. Seeing evidence of this online makes the experience all the more real for him. And being open on Facebook allows his friends and loved ones to celebrate this change in Adam, and to do it in a public way. This is social media functioning at its best—it's helping with Adam's self-esteem and making him feel more connected to and supported by his community.

"I got sixty 'likes' on that [post]," Adam tells me, referring to the change in his relationship status. "That's the most 'likes' I've ever gotten on a post." I ask Adam if he was excited about all the "likes," and he responds, "I was more excited about actually just beginning in the relationship than the Facebook 'likes.' That just was icing on the cake." Adam and his girlfriend became Facebook official nine months before our interview. Now they spend a lot of time messaging each other on Facebook and posting things on each other's walls.

And Adam always, *always* has her in his profile picture.

For many students, becoming Facebook official is a fun thing to do—though Adam might have been the happiest of all. In general, it doesn't involve much angst, and is just another step that people take as things get more serious. Like Adam, other people I interviewed were excited to go public about their relationships on social media, sharing a profile photo with a significant other and writing gooey things on each other's walls. Plenty of students were ambivalent about this change in their status—they didn't think much about it. They just did it. It's just what people do, in other words, so for many of them it's not a big deal. They have seen their friends do it, or they've done it before with previous boyfriends or girlfriends.

Generally, though, other people pay attention to shifts in relationship status and take notice when those profile pictures suddenly include a significant other. Many interviewees cited either their shift in status to "In a relationship" or a photo announcement of "We're officially together" as the post that got them the most "likes" ever on Facebook.

Another young man I interview speaks only briefly about being Facebook official with his girlfriend, but he does so nearly as enthusiastically as Adam. He'd mentioned his girlfriend earlier on in our conversation, so eventually I ask him if they are Facebook official.

"Yes," he says. "More so because I wanted the world to know that I was *so happy* to be with her. I guess, like, standing on top of a mountain and shouting it out to the world. That was my way of doing that, and I also wanted every other guy out there to know." He laughs at this part. "You mess with her, you mess with me." If you are in a relationship, he says, everyone starts asking you if you're being public about it on social media. "Is it Facebook official? That's what you hear all the time. Is it Facebook official?" he repeats. "Are you on Facebook?"

Jake, who attends an evangelical college in the Southwest, also speaks at length about how common it has become to make your relationship Facebook official, and how important it is to do it. "I had gotten in my relationship abroad and we started dating, and we became exclusive, and then we asked to be boyfriend and girlfriend," Jake says; here, he's talking about the real-life—as opposed to Facebook—evolution of his relationship. "And it was always that moment of, well, 'Let me tell my friends really quick first before we make it Facebook official,'" Jake goes on. "It's almost like our society has placed Facebook as an establishment.

For example, when people get engaged, that's how they announce it, on Facebook, and we've coined the term 'Facebook official' because now it's so official to announce through Facebook because everyone has a Facebook. It's *such* an official statement when you make it on Facebook."

This was not the case with Jake's previous relationship. He and his former girlfriend weren't public about it, and that didn't sit right with Jake. "For example, my last relationship, my ex didn't want to make it Facebook official, right off the bat, and that was something that made me wonder too. Like, well why *not*? I felt like, if we're going to share that, if you don't care that our friends know, then why would we care that everybody else knows as well?"

Jake is very happy that he and his current girlfriend are Facebook official. But it can have its downsides, the main one being that, if you're open about the existence of your relationship on social media, there is a possibility that any future breakup will be public as well. "That's one thing that really makes it clear to the rest, too," Jake says. Jake's glad that Facebook allows you to prevent your status change from showing up on the feed. "When you go back to single, I think it changes your status, but I don't think it posts that you went from in a relationship to single," he explains. "So I would say that it's visible to those who would want to go look; however, I don't think it's advertised in the same way that it would be if you got into a relationship and they put that post out there."

That's a relief.

Students tended to be laid back both about making their relationships public and about shifting their status back to "single" if they broke up. The comings and goings of various significant others is just another aspect of being on social media that young adults seem to be growing used to—it happens to everyone; therefore, if and when it happens to you, it's normal. You are not alone. The announcement of a new relationship is greeted with joy and approval (usually), and breakups for the most part are treated quietly and discreetly. People know that commenting publicly about someone else's breakup is something you just do not do, which makes most students feel a bit safer about not having to go through the end of a relationship in so public a way—though they realize that people will figure it out. The "hiding" of the breakup is another means of preserving the veneer of happiness and perfection.[1]

Being Facebook official is generally something students want to attach to their names.[2] It is a personal statement that can boost one's image. It says to everyone, "Somebody loves me and thinks I'm special," and it emphasizes that a person is "normal" and not a social outcast. And it's one of the key highlights on most people's highlight reels.

There are exceptions, however.

AINSLEY AND PETER: BREAKING UP IS EVEN HARDER TO DO ON FACEBOOK

Ainsley is unlike any other student whom I interview at her evangelical Christian college—or anywhere, for that matter. She's wearing *a lot* of makeup and is covered in jewelry. She is very nice, even sweet, but she talks like a Valley Girl, and on several occasions I feel as if I'm interviewing a reality TV star. Ainsley turns out to be something of a Baptist party girl. And she has had a lot of trouble with regard to her romantic life when it comes to Facebook.

Ainsley is very active on social media, and her romantic troubles on Facebook started early.

Ainsley has a nosy mother, who found out through social media that her daughter was having sex. One day Ainsley accidentally left her Facebook page open on her computer, and her mother read all her private messages—not her public updates. "I got caught for so much," Ainsley says. "I was grounded for a whole semester of school." Her parents are very conservative Baptists, and their daughter was doing many things that they did not like one bit. "[My mother] found out that I had had sex, that I had drank, and that I had done drugs, too. So it was just, three big things that she just found out about." Things were bad between Ainsley and her parents for a long time after this—she didn't trust them, and they didn't trust her. Things have calmed down since then, but the experience changed Ainsley's relationship to social media.

Today, "if it's ugly, I don't post it," Ainsley says, expressing the rule I hear from just about everyone. No cursing, no negativity, because "so many people, like, business people or people who are trying to hire you for a job, they're going to go, and they're going to look at [your social media.]" Most of all, Ainsley doesn't "want to come across as being sad.

And depressed," which means that things get difficult when she goes through a breakup. She admits that at one point in her life she tweeted "sad song lyrics," but then she quickly deleted them because "I don't want to have sad Tweets," she says. "I don't want people to know that I'm sad and feeling vulnerable."

I ask whether the relationship that made her sad was public online. "Yeah, I made it Facebook official," she says. "And, I don't know, I wish I wouldn't have, though. Because, when we broke up, he took it off of his Facebook, and that was, like, a sob in my heart and I don't know why. It hurt that he took it off of his Facebook." Ainsley's demeanor changes as she talks about this, from outgoing, outrageous reality TV star to pretty typical college student struggling over a breakup. The pain she experienced watching her breakup happen on Facebook, going through the steps of untangling their relationship in public, is evident on her face. In some ways, the Facebook breakup and the taking down (or, at least, untagging) of photos are akin to giving back someone's stuff—except that it happens in a way that all your other Facebook friends might notice.

Ainsley worried about people noticing. A lot.

"That's what I thought about the entire time," she tells me. "I was like, people are going to see this and think that I'm so sad. Like, wonder why we're broken up and ask me, and it's just going to be annoying. Plus I was worried about who was going to be 'liking' his, you know, [status], because on Facebook, it'll say, 'single' on his wall, and I want to see who's liking that. What girls are going on and are going to be like, 'Okay, he's single.' So that was definitely a struggle for me. I would go and check it all the time." It made her feel stressed, sad, jealous, and a thousand other unpleasant emotions.

Even before they broke up, being public about their relationship on social media caused all sorts of problems for the couple. "My boyfriend, he was very controlling," Ainsley says. "And, me and my friend from high school, he wanted me to stop being friends with her because I got into so much trouble with her. And so I told him, 'Okay, I'll stop,' because I was so into the relationship. And then, I would go and 'like' a post of hers on Instagram, and he would see that and get so mad at me. And so it definitely caused problems. And he would go and look at my Twitter from before we were dating, and like, the tweets that I did a long time ago that don't even matter, he would call me out on them. And I was like, 'We weren't even dating! It's just ridiculous. So it definitely caused a lot of annoyance in our relationship." Because of her boyfriend, Ainsley actually went back and did a Twitter cleanup of sorts, deleting some of her posts so they wouldn't bother him anymore.

This entire experience made Ainsley very cautious about being Facebook official in the future.

"I feel like, if I get into a relationship now, I don't think I'm going to put it on Facebook. I feel like it's so *public*. I'll post pictures on Instagram, if I get a boyfriend, of me and my boyfriend. But as far as making it Facebook official and having everybody . . . . Ainsley trails off, unsure how to finish this sentence. "I don't know. It's just, *my* business now."

With everyone worried about appearing perfect and happy on social media, going through a breakup on Facebook—changing your status away from "In a Relationship" to "Single"—can be complicated. There are ways to make the shift inconspicuous, but even then you may live in dread, wondering if people will notice. Dealing with whether or not to take down photos with one's ex not only is painful but also can make it difficult to maintain the appearance of happiness and perfection. The requirement to appear happy all the time is hard to fulfill when you are going through a tough time.

A painful breakup is also a reason some people quit their social media accounts for a while—if not permanently—simply so they don't have to broadcast the process of separation.

For example, Peter's girlfriend broke up with him, and being on Facebook, he says, causes him a lot of pain. "It does," he says. "In some sense just because there are still a lot of memories locked up in that social media. Because you build this profile, kind of stock it, you make it that representation of you and you try to make that authentic if you can. And for a long time I was the kid with the girlfriend, and you know, we have a lot of pictures up and that sort of thing. Occasionally, you know, I'll be running through something and looking for a picture from months ago, and I'll find a picture of us and it's like . . . . he trails off, unable to finish his sentence. Peter is still grieving months after the breakup.

Then Peter mentions that he has given up Facebook. I ask him how long it's been. "It was last night that I was just like, okay, you've got to take a break." The photos Peter and his girlfriend put up are still there. Neither one of them has made the move to take them down. "I don't know if it will come to that. It might if I end up back on Facebook in a few weeks, if I'll see them all gone or her tags gone or something." Peter didn't delete his account, he deactivated it—there's a big difference. If you delete your account, it's gone for good and you lose everything on it, but if you deactivate it, you can always go back and reactivate it again. "I did not [delete it]," he says. "When it comes to communication with some people, that's the only way I can communicate, so I don't want to destroy everything, the entire network that I've built. So I deactivated it, which is kind of an intermediary step."

I ask Peter how it feels to have deactivated his account.

"It's kind of nice just to not have that worry, not that I'm really concerned about Facebook," he says. "Just kind of not having that at all, just being able to enjoy life in the sense of being present, not that you can't enjoy life with social media because you very much can, but for some reason to me it's been more of a calming existence [without it]. Especially when I was off social media for a while—I did it after she broke up with me too, when I was bored I wouldn't check Twitter, I would sit and think and reflect and use that time for what some would say was much more constructive purposes than just seeing what my friends were doing on Twitter."

So this isn't the first time Peter has quit his accounts, but both times it was a breakup that prompted him to do so. Being on social media, for Peter, is difficult when things aren't great. "It's probably somewhat of a psychological thing, that when you're going through a difficult time and you want to talk to someone about it, I think humanity just in general, craves an interpersonal connection," Peter explains. "You can't get that through a computer screen, and I wouldn't say necessarily that those [social media] connections would cause pain as much as they wouldn't be too helpful to me."

Neither Peter nor his girlfriend posted about the breakup. "I don't think anyone else should see it like that," he says. "That's your personal life and really not anyone else's business. . . . This is going to seem harsh to a lot of people, but if you're going through a hard time or you're having a rough time, don't go to social media about it. Go talk to a person about it and deal with reality. Because social media is social media, and reality is reality."

EDDISON AND TARA: COMING OUT

The moment Eddison walks in the room, I know I'm in for a lively conversation. He wears a big smile on his face, and his dress is impeccable—unusual for a college student. He wears a bow tie, a sharp tweed jacket, a pink-and-white pinstriped oxford, and hipster glasses. Eddison is African American and also gay, but when I ask him about his sexual orientation (for demographic purposes) at the very beginning of the interview, he answers, "I am in between now."

Eventually I ask him what he meant by this.

"I know that I like guys, that's the thing, but back at home"—Eddison pauses a moment to think, then goes on—"Me being here [at school] makes my mom happy, and that's what I want to do. But also me being with a woman and having children makes her happy, so I've just been trying to find a way to officially say it. I spend a lot of time trying to make my family happy before myself, so that comes as a struggle. But here, I know what I want, I know what I want to do with my life. I know I don't want kids and stuff, but when I go home she's like, 'When are you going to get married? When are you going to have kids?' Yadda, yadda, yadda. And I say, I don't want to get married. I don't want to have kids. I'm going to get a dog. That's how I'm going to live. And just be happy with it. But it doesn't seem to work, so I've just been working on a better method."

Eddison comes from a big churchgoing family, and this is part of the problem. His mother thinks it's "not acceptable" to be gay, since it's not okay within their church. Because Eddison's mother doesn't know he's gay, even though he's obviously very close to her, it's a relief for him to be away at college. It gives him the freedom to be himself. But he still misses his mother and likes going home, too. "At home I'm still happy, I'm just more in my shell," he says.

That Eddison isn't out to his mother gets even more complicated on social media, in particular on Facebook, because Eddison is friends with her and several other family members. In fact, when Eddison became friends with his mother on Facebook—she required him to do it—he had to do a Facebook Cleanup, but not the kind that is done for professional purposes.

"She told me when I went off to college that we *had* to become Facebook friends, because we weren't at first," he says. "I guess that was a way to keep track of me. . . . She told me the night before I was leaving, 'All right, now you have to have me as a friend on Facebook.' So I had to go on Facebook and hide everything that I didn't want her to see, because there was a point when I was younger when I did post stuff about relationships. Or somebody like a guy that I liked. Facebook has notes—I don't know if it still has it—but a notes section where you can write notes and everybody can see them so I used to write poetry and stuff to this guy. He didn't know that it was about him, but it was a way for me to get it out there. It wasn't ever negative. So it was never, 'I hate you. You suck.' But it was more so love and positive. So there was a point in time where I did post stuff about relationships. And then once me and my mom became friends that's when I said ok, time to . . . . Eddison trails off.

Time to what? I ask.

"I think I hid it, but I could have deleted it because I can't find that function anymore," he says.

That was the end of social media romance for Eddison. Today he is a different person on Facebook than when he first joined. "I made [my Facebook] more about the professional aspect of who I am," he says. "I'd rather you see what I'm doing career-wise, what I'm doing, how I'm doing it, than to see something about a relationship or a breakup or stuff like that. When I ask if he ever posts about liking someone or being in a relationship, his eyes get wide. "No, no," he says. "When it comes to relationships and things like that, I am very, very private. I keep it all to myself. . . . I feel like when you make that public, you open it up to anybody's and everybody's opinion and interpretation of what it is. And I feel that my goal is to make sure that I make my family happy first. Putting something like that on Facebook: 'Yes, I am in a relationship with [a man],' I feel like that's unauthentic . . . .if I want people to know, Facebook shouldn't be my way to do it."

What becomes clear as Eddison continues talking is that his impulse toward privacy around relationships has more to do with the fact that posting about one would effectively be coming out on social media, as opposed to coming out in person. He doesn't want the people he cares about who don't know he's gay to find out through Facebook. He also doesn't want to mention getting into a relationship on social media—not even a comment about wanting to be in one—because of the way this could get interpreted (and misinterpreted) by different parties. "I don't want to lead people to false hope," Eddison says, meaning he doesn't want to lead his mother to think that he's looking for a woman to date. "I can say 'looking for a relationship,' and half of my Facebook friends interpret it as 'Oh, he's looking for a guy,' but then you have the other half, which includes my family, 'Oh, he's looking for a nice young lady to settle down with.' So I just avoid it altogether."

I ask Eddison if it makes him sad that he can't be honest about this side of himself on social media. "Well, no, that's not a sad thing," he says. "And as big of an influence as social media is starting to play with job searches and things like that, you just want to make sure you're covered. Not that being gay is a problem in the workplace, but you don't want that to be the first thing that someone sees about you, I don't know . . . . Eddison trails off after saying this. When he continues, what he says is punctuated by a lot of pauses and hesitations. "I don't want [people] to feel like that is who I am, it defines my whole person. So when you see that [so and so] is in a relationship with some guy, then it defines him. . . . Yes, gay pride . . . .gay, gay, gay, gay all the time. But most people think [being gay] may have a negative influence on work and stuff like that. So I just keep my personal life personal."

As Eddison continues to talk, several additional things become clear. He worries that coming out on social media—and having that go onto his permanent record—might jeopardize his ability to get a job in the future. Being gay is not something Eddison wants on his highlight reel. And though it seems okay for lots of people to be open about their romantic relationships on social media, he thinks it's really only okay for heterosexual couples. Eddison feels that to be open about a gay relationship is "extreme."

"I think it's extreme when *other* people look at it," he says. "So not necessarily from my point of view but from *other* people's point of view. Like, I can hear my mom saying, 'You are who you are and I may not accept it, but you didn't have to take it to that extreme of putting it on Facebook for everyone to see.' So, I think it becomes a big deal. People make it what they want to make it. A lot of people choose to make things like that a big deal. So I don't personally think it's extreme, it's just what other people interpret it as extreme."

So, his mother would think it's extreme?

"Oh, yeah," he agrees. "My mom would probably blow a couple fuses."

I ask Eddison if he thinks he'll eventually come out on Facebook.

He answers with a flat-out "No" and then sighs. "Right now I'm just trying to figure out a way for my mom to be okay with it."

Tara, like Eddison, is not out on social media. She is a junior who attends a Catholic university in the Northeast, and Tara thinks social media can hide things about a person, and therefore complicate a relationship—such as hide whether or not they're out.

Tara is currently in a very happy relationship with another young woman at her university. It's public among their friends, but definitely not on social media. Tara isn't out to her parents, and her girlfriend isn't out to hers either. Neither one of them is Facebook official about their relationship. Nowhere on any of their social media is there an statement that announces they're dating. "I mean, if you were to look, there's Twitter back and forth or like things like that," Tara says. "But there's nothing that says, 'In a Relationship with.' Nothing like that."

This is the first time Tara has been in a relationship with another woman, and she thinks social media just adds another layer of complication to coming out. I ask Tara if she and her girlfriend have ever had the DTR (define the relationship) talk that so many students have today about making things Facebook official, or if she thinks they will ever go public in this way. "I mean, it's more of a joking thing because we don't think that it needs to be public because there's so many people on Facebook that we wouldn't necessarily tell them, 'Oh, hey, by the way . . . .'" Tara says. "So we don't find the need to do it, or it's more of a joke when we talk about it than anything."

I ask Tara if it's just that she is a private person or if she worries about how people would react if they found out she's in a relationship with another woman. "A little bit of both," she answers. "I'm worried how my family would act, react. As for other people, I really don't care how they react. I'm not someone who's like, 'Oh no, so-and-so thinks it's not okay, so I'm going to hide it from them.' It's just I don't feel the need to publicly tell everyone, I guess."

I ask Tara if she thinks she'll ever come out online.

"Um, maybe, I guess, but I don't know," she responds, obviously uncertain. "Probably down the road, but not right now."

While many students are out on social media, many others are concerned about how openly identifying as LGBTQ might affect their professional futures and their social and familial lives. The possibility that adding "gay" or "lesbian" to one's highlight reel might cause them trouble is something they worry about. Eddison especially seems to be living out the social media divide: sticking to the "Everything is great!" veneer one is supposed to present online, while leaving off anything that might be perceived as "negative" or worrisome to others. What's sad, of course, is that the "negative" Eddison leaves out is his sexual orientation. He is hiding a core part of himself from a social media world in which he participates nearly constantly.

DINAH: FACEBOOK IS THE NEW MATCHMAKER

Dinah, the young Orthodox Jewish woman we met earlier who doesn't go on social media, has fascinating things to say about how social media is affecting the way Orthodox Jews find potential husbands and wives, taking the notion of making your relationship status "official" to a new level.

But first, she says, she needs to explain what Orthodox Jews believe about dating in general. "As a rule, religious Jews date to marry. We don't date for recreation," she tells me. "Most [non-Orthodox] people, from what I've seen, have a boyfriend to have a boyfriend. Or have a girlfriend to have a girlfriend. And we don't do that. So I will date someone as dating not as hanging out, [but] to see if he may be the one I want to marry and for him to see if I might be the one that he wants to marry. It may be set up by my father, it may be set up by a friend. If I met someone who I wanted to date, I could, but my father is my father, and I would have to tell him in advance. So I haven't had any boyfriends because even if I were dating someone, we don't consider them boyfriends—they are the boy you're dating until you get engaged. And then they're your fiancé or your *chossen*, which is Hebrew for groom."

Orthodox Jews find each other online today just like everybody else does, though there are a few significant differences. On Jewish dating websites, according to Dinah, "You submit what's called a résumé, and it details what you're looking for in a guy, what your background is, what your family's background is." The biggest difference? After you submit the résumé, it isn't the computer that does the matching for you. There isn't some online algorithm that puts people together. *Actual people* are behind the computers. "They're called *shad-chanim*, which are basically matchmakers," Dinah says. "They match up the résumés and decide, this person should go out with this person and this person should go out with this person. There are real-life *shad-chanim* also. You can go to a *shadchan*, submit your résumé, have them talk to you, and then have them set you up with someone. But that it's online now is interesting."

One of the notable things about going to an online *shadchan*, Dinah tells me, is that you can get matched with someone who's not in your own community. The person could be anywhere, which opens up many new possibilities. "And you can specify, I think, how far you want to go," how far you are willing to travel to find a possible match. And while this widens the pool of possible matches, distance can be a challenge. Dinah almost went out with a boy who lived four hours from her house. "We went to go meet, and the whole time I was thinking, so if we end up going out will he come here? . . . .The girl goes to the boy first, if it's far enough away, and they go out once or twice depending on the length of the stay. And then the boy will come up, say, for a weekend, and go out with the girl, like, two or three times, and then he'll go back home, and the girl will come and do the same thing. It's really difficult and really strange. And if you end up getting engaged, then there's the whole, well, 'Where are we going to hold the engagement party?' The wedding is usually held where the girl lives. And there may be multiple engagement parties, it's just difficult, so a lot of people don't want to do [long-distance dating]."

I ask if the Internet and social media platforms like Facebook help ease the strain of distance. "Traditionally you don't communicate with the person you're going out with," Dinah says. "Not to say it never happens. Usually they date frequently enough that it's not that weird." Also, Dinah explains, there is the miracle of Skype. If the couple lives far enough away, they may have Skype dates—approved and chaperoned—but not the traditionalists, she says. They don't even allow people from different towns to date. "So if you're dating an out-of-town boy, chances are your family isn't going to really object to Skype dating. But then Skype dates are like regular dates. You don't just decide, 'Oh, I want to talk to him, let's Skype him.' You make an appointment, and you talk about dating-type stuff, for say two or three hours, and then you hang up. So it's like going on a real date but minus the food."

Remembering how Alima's Muslim faith shapes her behavior on social media, I ask Dinah whether Facebook and other platforms are similarly complicated for young, unmarried Orthodox Jews. How does one navigate online "friendships" with someone you might like to date? Are boys and girls outside each other's families even allowed to be "friends" on Facebook?

"Typically, it's discouraged, [but] most people I know probably don't care," Dinah says. "And if they really think their parents are going to have a problem with it, then they block their parents from their Facebook account." She pauses, then tries to explain. "Okay, let's put it this way: if you're friending a boy on Facebook, you're probably also talking to boys in real life. It's highly unlikely that you're going to friend a boy if you aren't speaking to them. [At a traditionalist university] where they have different times for the boys and the girls to have classes, you technically go to school with boys, but they're not in your classes, so you have no reason to know anyone and you don't friend them on Facebook and you don't know they exist." At a public university like the one Dinah attends, however, everything is different. "You may be in a class with two other Jews, and one of them is a boy or both of them are boys. You're not going to ignore each other usually because they're boys. Here, I've been in classes with boys, I've been lab partners with boys, and it wouldn't be strange if I friended a boy on Facebook." Dinah laughs. "And my father's understanding of social media is nonexistent." She laughs again and says, "If he found out I friended a boy on Facebook, he would probably freak out, and I would have to explain it to him, and he still wouldn't really get it, but he would probably calm down while I tell him, 'Well, everyone does this, it isn't so weird, I talk to him in school.'"

Once you have that friend on Facebook, though, flirting and falling for one another become temptations. Some girls Dinah knows don't ever friend boys on Facebook, for exactly this reason. "It's kind of a little on the scandalous side to go out with a boy you met, rather than a boy you were set up with," she explains. "There's actually a known entity of people who will go to a *shadchan* and say, 'Hi, my daughter wants to date this boy, and can you be the *shadchan* for them?' It's a little weird. It happens quite often, but it's kept very quiet."

For Orthodox Jewish parents who've ended up in this predicament— they have a son or daughter who met a person and then struck up a relationship of sorts online. Now their children want to date each other. Going to a *shadchan* to request that this person make official a match that the young people have already made with each other *unofficially* is a means of saving face. It's one way the Orthodox community is coping with the fact that social media makes flirting and connecting easy for young, unmarried men and women *outside* the boundaries of tradition and their families.

"I have a friend who actually wants to do this," Dinah admits. "She met a boy here [at school], she was in classes with him, she friended him on Facebook. Her parents did not know." Dinah sighs. "Not that she was actively hiding it from them, but she didn't hide it any more than she hides everything else.... She finally told her mother about two weeks ago ... and so [in theory] her mother would then go to a *shadchan* and say, 'Would you set these two up?' And then they would go out dating officially. I don't know if it's going to happen [with my friend], but it *does* happen. This is what happens when you go on Facebook, and when you go to school with boys."

I ask Dinah if she thinks that two young people dating because "they met," not because "they were set up" by their parents via a matchmaker, happens more now because of social media. Yes, she tells me. Before, "it was very hard for girls to meet boys.... If you saw your neighbor in the supermarket, you weren't allowed to talk to him," Dinah says. "Now people are going to college and all of a sudden they're meeting boys .... [and] now it's much easier to stay in touch with them because no one's watching over your shoulder while you're online." She pauses, reflecting for a moment. "So I think that online definitely has a lot to do with boys and girls communicating more, but it's just easier to be secret about it." Dinah backtracks and explains that there have always been boys and girls who've met and fallen in love in secret. But the Internet and social media make it far easier to do this today. "The Internet makes everything else easier to do. It's easier to find recipes, it's easier to find songs, it's easier to find other people who have the same crazy chronic medical conditions as you do. It's easy to find everything, so of course it's easier to find boys. It's a side effect of the Internet," Dinah offers. "Which a lot of people don't like, and that's why there are some people who are Orthodox who don't use Facebook."

One last time, I decide to press Dinah about whether this has anything to do with her own shunning of social media.

Dinah grins. "Nope," she says.

HANNAH: THE FACEBOOK (RELATIONSHIP) CLEANUP

While Ainsley mentioned in passing that she did a Twitter cleanup to get her overbearing boyfriend to stop bothering her about past tweets, Hannah, the "manicurist," tells me about an alternate version of the Facebook Cleanup. Hannah didn't want to make her current relationship public, precisely to avoid some of the problems cited by Ainsley and Peter. Like Ainsley, she's been burned before. And she had a long conversation with her current boyfriend when the topic of making their relationship Facebook official came up.

"I said, 'Look, it's difficult for me to put this relationship status with you on Facebook because if we were to break up, everybody would be in my business, and if we were to break up, that would be a time when I needed to be alone, and needed to be introspective," Hannah says. "And I've had failed relationships in the past that have been on Facebook, so I don't need people knowing my business. I think it was a source of tension, because he was like, 'If you really love me, you'll do it because I don't want you to be embarrassed of me.'"

Hannah eventually relented, deciding it's *probably* not a big deal, since she plans on being with her current boyfriend for a long time. She feels more serious about him than she's ever felt about anyone. "He sort of talked me through it. But now I kind of like it because I like posting pictures of us that I like together."

But it hasn't been all roses either. "It's definitely caused conflict," Hannah tells me. "If I had pictures up still from another guy that I was seeing, he was like, 'Can you please take those down?' He was going through my entire Facebook, so I was like, 'Well, I'm going to go through yours if you're doing that to mine.' So he had this long post about this girl he had dated, like, 'I love you, I'm so glad we've been together for this long,' blah blah blah. Then he had song lyrics to a song that he was, like, 'This is our [his and Hannah's] song,' but he had the song lyrics with *that* girl [too], and I was just like, 'But you used that song with me! You can't use the song with both of us!'" Both Hannah and her boyfriend saw all the different women and men each other had "liked" in the past, which caused conflict and hurt feelings, as well as raising questions about *why* each person had chosen to "like" a certain post or photo.

It was these sorts of unpleasant discoveries that prompted their own version of the Facebook Cleanup, so they could make their relationship look like the only relationship either one of them had ever had.

"We definitely went through every single social media post," Hannah says. "And we're like, 'Delete that. Delete that. Delete *that*,' because we wanted to have this image that we were together." They did this together, but Hannah insists that it was all her boyfriend's idea.

ON FRIENDSHIP: COLLEGE IS SO PERSONAL.

With each student I interviewed, one of the first questions I asked was, "How did you meet your friends at college?" I brought up this subject prior to any conversation about students' online lives because I wondered whether students would tell me that they met their friends online. But out of nearly two hundred interviews, only a handful of students told me that they met their college friends through Facebook or other social media platforms. Most often, using social media was a way of getting to know their roommate before they arrived on campus, or checking out a Facebook page created to give soon-to-arrive first-year students a place to connect. But finding an actual friend this way, especially after arriving on campus, was exceedingly rare.

College students do *not* meet their friends through social media. They meet them in person—in dorms, in classes, through activities, or through other friends—and they like it that way. In other words, the ways in which college students meet each other on campus are the same ways that college students have always met each other on campus. Social media exists, sure, and someone may meet a potential friend in person and then log on to check that person out. Social media certainly facilitates making plans, but it is rarely the starting point for a new friendship.

For the students I spoke with, the idea that they would initiate contact and get to know a potential friend via social media—without having already met them in person—was unpleasant, uncomfortable, even icky and weird. They are not interested in meeting people on social media. Social media is a useful tool for maintaining friendships (especially long-distance ones) and making further contact, but not for an introduction to someone new. Students use social media as a convenient way to send group messages and announce a party, for chatting about class assignments, or for deciding where to meet up later. This aside, the college students I spoke with were delighted that being on a college campus offers them a physical, interpersonal space where the possibility of meeting friends is everywhere you turn. They worry that—because of social media—soon these face-to-face meetings might go away.

I mention this for two reasons. First, many adults today seem to think that the generations coming up no longer have interpersonal skills. They assume that young people interact only through social media and aren't interested in face-to-face interactions. Second, and most important, there is so much speculation that the college campus experience of the future (or at least a lot of it) will occur online. While one can make both a convenience argument and an economic argument in favor of online courses, the more we "upload," the more we are taking away from young adults who are at college to learn, yes, but also to live.

It may seem like they're online all the time, but they do not want to *live* there.

At its best, social media functions as a tool for navigating one's relationships. It is not a replacement for those relationships.[3] Relationships still happen in person—and young adults prefer it that way. They don't want social media to go away, but they *really* don't want the real world of face-to-face interactions to go away.

Students were split almost evenly between those who felt that social media—because it is such a great tool for keeping in touch—makes friendships "easier to maintain" over the long term, and those who felt that social media is making all of our relationships more "shallow" and "superficial." Some were simply thrilled that social media allows everyone to stay connected to each other, especially friends and loved ones who live in faraway places. Right alongside them were those who mourn the way social media allows everyone to "pretend" they have tons of really great friends, yet who believe that the quality of those friendships has plummeted dramatically, robbing them of the face-to-face joys and responsibilities that go along with true friendship. Even students who expressed their approval of the way social media allows people to keep in contact had complaints about the downside of relationships on social media: that friendships "need to be verified" publicly, and that just because friendships are "numerous" doesn't mean they have any depth or real commitment. Some lament that nobody really talks anymore, and that "friend has changed as a word" because of Facebook.

It's not only friendships, either, that college students still prefer to strike up in person. When it comes to dating, romance, sex, and hooking up, their preference for an in-person, face-to-face spark is undeniable.

9

THE ETHICS OF SEXTING

Tinder, Dating, and the Promise of Mutually Assured Destruction

Sexting responsibly would mean,
if you're not sending your pictures or,
like, sexy text messages to somebody
you don't know.

Jeremy, sophomore,
Catholic university

I do not know of any peers who engage
in [sexting] now, as it is widely known
to negatively impact your reputation on
campus. We are now adults as well, and
many more worry about their futures,
and whether or not future employers will
see these images. At least, I do.

Erin, junior,
evangelical Christian college

LAUREN: A TOTAL DATING FAIL

Lauren, a sophomore at a public university, tried to use Facebook to spark a romantic connection with a guy she's interested in. He's in one of her classes, but they've never actually spoken.

"There's this guy in sociology class—he's beautiful," she begins. "He's a basketball player, and the cutest thing ever! But, yeah, I like, creeped on him on Facebook and Twitter to see if he had a girlfriend. He doesn't, but I've also tried to talk to him on there [on Facebook], and he hasn't said anything to me, so I'm just like, eh," she adds with a shrug. College students talk a lot about "creeping" on people, which means gathering information about people from their social media accounts. It can apply to friends and acquaintances and even people you've never met, but when students creep on someone, it's usually someone in whom they have a romantic interest.

I ask Lauren what she means by "trying to talk to him on there."

"I tried to contact him and be like, 'Hey,' because we had a big exam," she explains. "And I messaged him on there, and I was like, 'Hey, how'd you do on the exam?' like, 'Do you think you did well?' and just tried to spark up a conversation, and he never responded." Lauren sighs here. "So I was like, 'That is *so* awkward'!" It turns out that this happened just a few days before our interview. Lauren seems pretty disappointed that her efforts failed, so I try to console her a bit.

"Well, basketball season is just starting, so maybe he's busy," I tell her.

She laughs and shakes her head. "No."

"Are you *sure* he saw it?" I ask.

"I'm *positive* he saw it," she answers.

"Did you get the little indicator that said he opened the message?"

"Yeah," she says. "It said, 'Seen,' and that's, well, that's *awkward*, but that's okay. I don't care."

"Have you ever successfully met someone on social media?" I ask next.

"No," Lauren says.

I find myself wanting to convince Lauren that all is not lost, that just because this experiment in Facebook romance didn't work out, she might have better luck next time. "So that was your first try?"

Lauren nods. "That was my first attempt and my first fail, and then I'm not going to do it *ever* again."

Trying to find a date via social media, it seems, is not Lauren's thing.

Online dating and hooking up are subjects of constant media fascination. Because I have written about sex on campus, I am frequently asked about how college students are using social media to date and hook up. The media can't resist this combination of two topics—sex and technology—with which our culture is obsessed.

So, how *are* college students using social media to meet potential partners? Here's the truth: they aren't. At least not all that much. I want to be clear: just as there is a difference for students between *meeting* new friends via social media and using social media as a *tool* to conduct and maintain friendships, the same distinction applies here. You might use social media to negotiate the general logistics of meeting up, or for flirting. But I traveled far and wide, to every kind of college, and it was difficult to find college students who used social media as a first-stop tool for dating, or even for hooking up.

Lauren is rare in that she even tried.

There was that rather geeky young man at an evangelical Christian university who told me he'd tried to use Facebook as a way to meet girls. His method was to message them with the line, "Hey, so you wanna be Facebook official?" But all his efforts ended in massive failure—people didn't even think it was funny. So, like Lauren, he was over it.

I met a few students who had tried online dating, or who knew people who'd tried it. But it's extremely uncommon for students to strike up a romantic connection online if the student has never met his or her romantic interest before.

For instance, when I ask Brandy about online dating apps like Tinder, she is highly skeptical.

"I like meeting people organically," Brandy says. "I like seeing them face to face, reading their body language. For me that face to face is really important because I feel like that's the best way you can really read a person and see where they're at. . . . I don't personally have any of those [apps]. I got too many boys in my yard already." Brandy laughs. "I don't know if I could *ever* just not meet somebody face to face and have them approach me in a romantic way."

Brandy has seen some of her friends take the plunge into online dating, but she does not approve. "So it'll be interesting [in the future] just because I see a lot of people moving more to an online space. Like, one of my friends just got a boyfriend. This is the dumbest thing, let me tell you. She literally met this guy on OkCupid and literally a week after that, they started dating in an exclusive relationship. That blew my mind because I'm like, 'You've literally known him for like a minute.' And, like, he could be a serial killer!"

In fact, when students *did* discuss the subject of online dating, it was generally with reference to those weird, reckless adults (some of whom are their single parents) who actually go out with strangers they've met online. It's something that young people have trouble fathoming. After all, they've been taught their whole lives that it's *dangerous* to meet up with people you've only chatted with online.

Most were dismayed by the whole phenomenon. The average college student dreads the possibility of one day being so desperate that they might have to try it. Campus provides them with a universe of potential dates (or, more likely, hookups) they can meet in person, and they would prefer to not have this aspect of their lives usurped by social media.

THE PROS AND CONS OF TINDER

There are a number of dating apps out there, though they are more commonly thought of as hookup apps. These include Clover, Hinge, and Grindr, but by far the most popular (as of this writing) is Tinder. For the uninitiated, Tinder uses the GPS on a person's phone to pull his or her location. You set it up so that the app knows who you are looking for (by gender, age, etc.) and the radius in which you are looking (one mile? ten?). Once Tinder knows your parameters, photos (with brief tag lines) start appearing on your screen. It's like a candy store of potential dates—you go through them one after another, swiping left if you're not interested, right if you are. When you swipe right and then "heart" someone, if that person has also "hearted" you, you get a "match," and you're able to get in touch with that person—for a date, in theory, though it's famously used for easy sex and hookups. *Vanity Fair* writer Nancy Jo Sales published a lengthy article in August 2014 about how apps like Tinder are changing—destroying, in her view—dating as we know it, essentially replacing it entirely with hookups. The company behind Tinder responded on Twitter, quite viciously, accusing Sales of ignoring the benefits of the app.[1]

Tinder is definitely present on college campuses, though it's not as common as one might think. Some students are on it and enjoy it, but more often than not, if the subject of Tinder came up during interviews, the students looked disgusted and went on to tell me how much they dislike the whole business.

Joy, a junior at a private university, let her feelings be known—bluntly. "Honestly, I think it's a little ridiculous," she says, laughing. "Tinder to me is like, 'What the fuck are you doing?' And people are in relationships from Tinder now, and to me, I wouldn't even want to pursue a meeting. I wouldn't want to download Tinder because I would never want to have to tell someone, 'I met him on Tinder.'"

In other words, if you can't meet people in person, especially in college, you've failed. And you should be embarrassed. Joy goes on. "That's the last thing I would ever want to have to say, and I know people who do [meet on Tinder]. This one girl was in my media class, and we were talking about Tinder, and I was saying how ridiculous it was, and how someone from my house I had heard talking about how she was going to go on a date with this guy from Tinder, and I was like, 'That's really creepy,' and then a girl in my class was like, 'I met my boyfriend on Tinder,' and I was like, 'Oh, I'm sorry.'"

Joy shakes her head in dismay. In her view, joining Tinder is a sign of desperation. And college students are just too young to get that desperate, in her opinion. "We wouldn't get on eHarmony and start trying to meet people at twenty years old, so why are we suddenly doing it this way [on Tinder]?"

Another young woman, Sage, from a Catholic university, had a more nuanced view. She doesn't have Tinder herself, but some of her friends do. "I've had different friends who have had different experiences with Tinder," she tells me. "My roommate freshman year got Tinder, and she met up with five guys, five *different* guys, from Tinder, and by 'met up,' I mean, this random guy she met on this social media website picked her up at the residence hall and they went and hooked up, and then she had one person come and like sleep over, just a random guy that she met on Tinder! So like, she *actually* used it for hooking up. But then one of my close friends at home thinks it's absolutely hilarious. She thinks it's so funny to just go through and look at the people. I've sat there with her, and we both just scroll through and people message. Guys will message her and say stupid pickup lines that are funny. So on the scale of things, I've seen people just use it kind of as a joke, and then people will actually use it to hook up with people."

This was literally the only time, in all my interviews, that anyone mentioned someone using Tinder to hook up with a stranger. As a rule, college students simply do not stoop (as Joy might put it) to consorting with strangers. Mostly, if students talked about using Tinder, it was as amusement, scrolling through profiles and laughing.

In the online survey, students were asked to name all the social media platforms they use on a regular basis. Of the students who answered this question, only 9 percent said they use Tinder regularly.

Because Tinder uses GPS, you can pretty much limit your choices to people on campus, and that's what the few students who use it do. They use it to flirt. Say there's a cute guy in your physics class but you've never actually met him? Maybe he shows up on Tinder when you're playing around on it some Friday night with your friends. This allows you to swipe right on his photo and "heart" him—and hope that maybe he's already done the same for you. Either way, voila! Once you swipe right and push that "heart" button, you've let him know you might be interested.

Flirting accomplished.

Maybe nothing happens from there—maybe he never responds, maybe he does, but you never talk to him in person. Or *maybe* next time you see him in physics class you actually have a conversation because you've established a connection on Tinder. Tinder can provide an opening to talk to someone you've always thought was attractive. Students find it incredibly difficult to establish that opening these days—going up to someone on campus they already find attractive and saying hello, in person, boggles their minds. Of course, once a connection is established on Tinder, if it leads to anything, it will likely be a hookup, not a date. Hookup culture dominates campuses. Dating (at least of the more traditional sort) is nearly nonexistent, even if students would prefer that not to be the case. It's definitely true that college students don't know how to date anymore. It's also true that most college students would *like* to date if they could. Tinder can help alleviate their fears and anxieties around that initial meeting (though it doesn't always function this way).

But I want to emphasize that the reputation Tinder has in the media and the fears stoked by alarmists—that Tinder simply facilitates sex between strangers—do not seem to apply much on college campuses. For those who do occasionally use Tinder to find hookups, they are almost always hookups with other students. Moreover, for college students, hookups are a broad category—they can be anything from kissing (and it is often just kissing) to sex. So even if a student uses Tinder to spark a hookup, that hookup may simply lead to an evening of making out with another student.[2]

The same dread that college students feel about online dating—the sense that meeting someone with whom you have *no prior real-life connection* is reckless—applies to Tinder also. Students may indeed want to have sex and hook up, but they do not want to do so with anonymous strangers. They want to have sex and hook up with that hot guy from American lit or that hot girl from chemistry class. Even if they have no prior formal introduction or relationship with that person, the very fact that this person is a known quantity—attends the same college, is in the same class, maybe even has some mutual friends—changes the dynamic entirely. For better or worse (and I would say for better overall), this makes the person with whom you are flirting and with whom you might like to meet up "safer." They are "safer" in the eyes of students because you are going to see them again in class, because you can get a sense of their reputation from others before anything happens between you, because you likely already know where they live or can find out easily, and because you will have further access to them if need be, since they live and go to school on your campus and are bound by its rules and authorities. Granted, this is not a guarantee that a hookup will turn out well, and it's *certainly* not a guarantee against sexual assault. But, despite fears expressed in the media, students almost never use Tinder to meet total strangers.

Tinder, for those who are on it, is a useful tool for showing interest, possibly for flirting, and definitely for a quick ego boost if someone needs one. Is Tinder a part of hookup culture on campus? Definitely. It depends on the campus, since Tinder is more popular on some campuses than others. And hookup culture was dominant on college campuses long before Tinder was invented. Hookups happen regardless of apps and social media, so while social media may play a role in hookup culture, it certainly didn't create it. And if social media were to disappear tomorrow, the effect on hookup culture would be pretty much nonexistent. Hookup culture would continue on, unhindered.

When I asked students about whether dating apps are changing the way people date, hook up, and have sex, only a small number (16.5 percent) mentioned Tinder. Of those who did, their feelings about the app were decidedly mixed, as apparent from the following:

Student Reply about Tinder	Percentage Out of 100
Tinder is definitely a hookup app	23%
Tinder is for flirting/fun/it's positive	18%
Tinder is both good and bad	1%
Tinder is a terrible thing	33%
Vague idea/no experience of Tinder	23%

Nearly a quarter of the students who commented on Tinder either had only a vague sense of its existence (they'd heard of people who used it) or simply wanted to point out that they'd never used it and didn't really know much about it. (Granted, things change so quickly online that Tinder's popularity, too, may change by the time this book sees the light of day—it might come to dominate students' romantic encounters, or it might fade from relevance altogether.) And nearly a quarter of the students said, in effect, "Of course Tinder is used for hooking up!" but didn't have a strong opinion about it. "It is an efficient platform for the casual hookup culture," one said. Period. But a small subset of students love Tinder (that 18 percent) and felt that these apps "can be helpful," especially if you're shy, and can really "open up your options." And about a third of the respondents *loathed* it.

"It's pretty neat to have someone matched on Tinder and then see them at a party later—you know there's at least some base level of interest," said one male first-year. "I think it just makes people more honest and clears up the fog of war, so to speak." One young woman, also a first-year at an evangelical Christian college, met her boyfriend through Tinder. "Most of the time, [Tinder] is seen and talked about as a joke and something that is not taken seriously," she wrote. "My experience, however, was positive and very different. I met my current boyfriend on Tinder. We have been in a very committed relationship for over one year now." Another young woman, a sophomore at her Catholic university, who identifies as gay (and checked "transgender" in the online survey), thought Tinder was especially useful for women who identify as lesbian. "Tinder is interesting when listed as a queer woman because lesbians essentially use it just for dating (legitimate, romantic dating)," she wrote. "While heterosexual couples appear to use it a lot more for hooking-up."

A number of students also used this question as an opportunity to comment on how Tinder didn't create hooking up. In fact, apps like Tinder help weed out the people who aren't into hooking up. "It's not like hookup apps invented hooking up," said a sophomore woman at a Catholic university. "There's always going to be some college kids who want casual sex, and it's good that there's a place for them to find each other without them having to risk making classmates and coworkers uncomfortable unnecessarily."

But most students took a practical attitude. In the words of a young woman at an evangelical Christian college, "It allows for very causal dating relationships. It also takes away the pressure of pursuit and rejection—people meet based on mutual attraction. This has its pros and cons. I think for the hookup culture, it has provided a lot more liberty. However, I think it has been detrimental to those who value long-lasting relationships based off more than physical attraction. It also creates the idea of dating as choosing from a pool rather than an individual connection."

Some students felt that it's all in how you use Tinder. "This is just a new way to date," said a young woman, a junior at her Catholic university. "Some are worse than others, but it depends on your intention. If you intend to go on Tinder to find someone to hook up with, then that's your choice but if it's your intention to find new and interesting people to form a relationship with, then that's your choice."

Then there were the students who rued the existence of apps like Tinder. "It is so negative and it is desensitizing," said a sophomore woman from an evangelical college. "My roommate uses Tinder and I disagree with it, because it takes away the idea of human interaction. I firmly believe that people need to build relationships face-to-face." A number of students who disapproved made comments similar to the young woman, a junior at an evangelical Christian college, who said, "Tinder I believe is especially bad, as it promotes superficiality in my generation, and encourages hookups," or this first-year woman at a private university, who said, "I think Tinder turns relationships and sex into something almost marketable—easily accessible online, and quickly discarded once 'used.'"

But, in answer to this one essay question in the online survey, many more students chose to comment generally about all online dating apps, which include Tinder, Grindr, and even Snapchat (in some students' opinions), without naming Tinder specifically. In fact, this optional essay question was one of the most popular in the entire survey. What's most interesting is that, though very few students claim to use these apps, they certainly have opinions about them.

The vast majority of those opinions—like students' overall opinions about online dating—are extremely negative. Young adults do *not* like the idea of meeting people they might date or hook up with online. Nor do they take advantage of any platform that widens their options beyond campus. And they are incredibly judgmental about those who do. What's more, in these particular answers, students' comments reveal a lot of negativity about hookup culture in general—and the reason students feel so strongly about these particular apps is because they believe they perpetuate hookup culture or, worse, exacerbate it and contribute to the devaluation of dating and sex on campus.

One young man said, simply and succinctly, that because of these types of apps, "hookup culture is becoming pervasive and commitments are being thrown out the door." A young woman commented, "It's not real and I think it makes it harder to find real relationships. I think it's totally changing the way people date, often promoting hooking up and sex instead of finding someone who genuinely cares about you."

Those students who offered an opinion on how these apps influence dating and sex on campus displayed severe distaste for them and the power they have to worsen what these students already see as a problematic dating and hookup scene. They frequently used words like "pathetic," "stupid," "sexist," "superficial," "terrible," and "horrible." They felt such apps could even be "dangerous" or lead to dangerous situations, and that they exacerbate the objectification of women or of people in general. These apps are disrespectful of sex, commitment, and human contact and have a detrimental effect on dating and sex.

SEXTING: A GENERATION DIVIDED

Another topic, besides so-called hookup apps, seems to cause otherwise reasonable people to panic: sexting. Generally, when I asked college students what they thought about sexting, I got one of two answers. One was a passionate *"No way! I would never do that!"* The other was a shrug. About half the students I interviewed thought that the people who sext are unbelievably stupid. The other half thought those people were lying because sexting is very common.[3]

Of course, students have very different ideas of what "sexting" actually means.

"I think it needs to be more defined, what sexting is," Mark begins. "Anything of a sexual manner, even if they're naked or not—I would consider that sexting. I know some people maybe would consider, if they're naked, or if they don't have certain clothing, then *that's* sexting. Or to where you're texting in kind of a sexual way. I think people try to make it more severe than what it really is. It's pretty wide, I think, how it should be defined." Mark goes so far as to include flirty texts as sexting, too. "Just like flirting, like, 'Oh, what are you wearing?' or 'What are we going to do later tonight?' Or sending pictures, like, if a girl was pulling her shirt down a little bit. Anything that has kind of a sexual reference I would consider sexting, if it's sending pictures, just texting or even if it's emailing."

Occasionally, students even interpreted "sexting" as using apps like Tinder or seeking sex via social media. Most understood the term as having to do with making explicit sexual remarks via text, and most commonly as sending suggestive or nude photos. Snapchat came up often in the context of discussions about sexting, since a lot of people think of Snapchat as the ideal sexting app.

Occasionally, someone like Matthew grew a little embarrassed discussing the topic. When I asked him about sexting, he was one of the students who immediately referenced Tinder. He felt like people on his campus used Tinder "all the time" to search for hookups—but, again, with other students, not with strangers. "If they're on Tinder, they're looking for somebody to hook up with that's like really close to them," he says with a laugh. They'll use Tinder to "invite them to the party that we're going to that night."

With regard to "sexting" in a more technical sense, Matthew thinks that it also "happens all the time." He likens it to phone sex of the past. Then he goes on to say something I find really interesting: "I'm actually guilty of [sexting] because this summer I sexted my girlfriend a few times because she lives three hours away from me so I can't be there all the time." I ask him why he said he was "guilty" of sexting. He chuckles, grows a little red, and starts stumbling over his words. "Well, I don't know, because I know that, people have, like a bad image of people, um, anyway, like my *parents* wouldn't approve and, like, um, like my *pastor* wouldn't approve. So, like, *guilty* is one of like my first reactions when I'm like talking to someone about it that's not my girlfriend, you know?" he says, laughing nervously again. I press him to consider why he feels embarrassed to admit this. "I shouldn't be doing [it] so I feel a little embarrassed about it. But, yeah, I've done it before because I can't be there all the time and it's just a way to, you know, keep a relationship intimate from a distance. So I don't think it's always a bad thing. I've only sexted in a relationship," he adds, laughing once more. "When people sext each other outside a relationship, I think it's very strange. Like, a few of the guys on the [baseball] team have sexted girls back home that they have not really anything to do with, and I don't know. I find that really weird. That's something I would never do, but I know it happens all the time."

Matthew draws a sharp distinction between sexting within a relationship and sexting outside a relationship. "I suppose you really should only [sext] with somebody you really trust and you're in a relationship with, like a *committed* relationship," he emphasizes. "Obviously breakups can go really bad, and if that person doesn't care about you as much as you thought, then maybe they'll do something to ruin your image."

Students who do sext occasionally often made it clear that they did so only within committed relationships. Sexting with a significant other is one thing; sexting with a stranger is something else entirely. Even students who said that they've never sexted, or never would, would often say that they thought it was okay for other people to sext with their significant others but never with a stranger. The latter is just too risky.

Vidya thinks that sexting is pretty universal among college students. "I know my friends do it," she says with a laugh. "Not that they've taken it too far, but it's definitely something that's a lot more common now than I feel like it was, especially with smartphones, especially things like Snapchat, where for ten seconds, you can see a photo, you know? And I know people do that.... But it's definitely a lot more common, especially with all these tools that you have, like front-facing cameras and FaceTime."

The idea that "sexting happens" just because we have the "tools" for it—front-facing cameras, apps like Snapchat, FaceTime—is the same argument a number of students made when talking about selfies. Technology makes sexting easy, so of course everyone does it.

One young man at a Catholic University had a rather romantic idea of why people shouldn't sext, and what sexting has to do with valuing (and devaluing) our bodies and relationships. "Sexting, it's not really something that somebody should be doing because it's disrespecting your body," he says. "It's kind of saying that all there is to your body is what you can see from the picture, whereas if you wait for somebody, like a significant other, to see the more private parts of your body until a later date like marriage or even, at least, farther into the relationship if not marriage, it's more special. It can be more romantic than just an outright naked picture."

Then there is the young woman who thinks sexting is just plain stupid: "I mean, you see people that are having scandals.... I mean, there's politicians that have been found with nude photos and stuff like that. Just leaving that footprint isn't a good idea. I don't know, maybe people get pleasure out of it, but is it really worth it if it's such a risk that someone could find those photos? And even if you're not a public figure, someone could use a photo they have of you to ruin your reputation even just within a college campus, within a high school, by distributing those photos. I mean I guess sending the photo is not really a big deal, but *I* would never do it."

Then there are students like Joy who have mixed feelings about sexting. Like many students, she feels that at a certain age, you're just too young for it and it's shocking, but as you get older it can be okay in the right circumstances. "I remember when I was in ninth grade and I had a phone, and some guy texted me asking for tit pics, and I was like, 'What?! No!' I wanted to cry. I was like, 'That's the most repulsive thing you're asking me for!' ....My philosophy on it used to be, when I was younger, that if you're not comfortable enough to see something in person, you shouldn't be seeing it on a screen." Joy feels you should be "old enough and mature enough" to do it, and she feels this way about herself now. She has a boyfriend, and sexting is a part of their relationship. "I'm not going to say I've never sent a sext, because if we're in a long-distance relationship, there's gonna be times where I might send you a picture of something, you know?" Joy still has limits on sexting, though. "I wouldn't do it as a casual [thing], with, you know, someone random," she adds.

Then there is Brenda. Unlike Joy, who grew *into* sexting as she got older, Brenda seems to have grown *out* of it. "It's common among people my age, not necessarily my friends," she says. "We talk about this all the time actually. I think a couple of us used to do it a little more frequently, including myself. It was a thing for us senior year of high school, but now we've grown out of it and it just really isn't a thing. . . . You can't just trust anyone with that kind of stuff."

Many people drew distinctions based on age and relationship status. And lots of the students I spoke with were worried about their younger siblings, who might not be mature enough to think through what they are doing. The consequences, they fear, could haunt them for life.

But no one had a more elaborate theory of sexting than Jeremy.

JEREMY: SEXTING THE RIGHT WAY

I meet Jeremy on a beautiful, sunny day at his idyllic Catholic college. He wears a colorful, tie-dyed bucket hat on his head, a Beatles T-shirt, skater shoes, and shorts, and he exudes a sort of relaxed enthusiasm. Jeremy is an incredibly fun person to interview—lively, hilarious at times, very intelligent and thoughtful. He's on just about every social media platform imaginable—Twitter, Tumblr, Snapchat, Vine, Tinder, Instagram, Facebook. He has a girlfriend back home, and when I ask him what sorts of things make him happy, he replies, "Too much makes me happy," without skipping a beat. He's a laid-back guy, disposed to enjoy life and have a positive attitude about things. In general, Jeremy posts whatever he wants on his accounts, though he does "watch his ass," he explains, because he doesn't want anybody Googling him and finding something for which he might get judged unfairly. He sees social media a just another outlet to express himself.

Jeremy enjoys selfies, he tells me, especially on Snapchat. But he prefers texting most of all because it's so private. Then I ask Jeremy what he thinks of sexting.

"I don't really have a problem with it," he begins. "Social media is way different now than it was, than it would be if it was presented a couple years ago. I think Snapchat and things like Tinder for sexting are more common in my generation, and I think we know how to use it. I mean, obviously not everybody, because there's people who abuse everything, but I think my generation knows how to sext in a responsible way, and they don't just send things out to random people. It's kinda like Snapchat is—Snapchat is connected to your contacts, so I know the people I'm sending it to. And Tinder, Tinder's kind of more of, like, a stranger meet-up site, but you talk to the person before. One of my buddies met his girlfriend on Tinder, and she's really cool, we all hang out with her. I like her, she's a nice person. So, I think our generation knows how to sext responsibly and not go meet somebody who might, you know, kidnap you."

Jeremy has mentioned "sexting responsibly" twice now, and I want to understand better what, exactly, he means.

At first, he tells me I've asked him a difficult question. But then he begins to parse out an answer. "Sexting responsibly," he says, pausing for another beat, "would mean, if you're not sending your pictures or, like, sexy text messages to somebody you don't know. I mean, I've sexted before, but it's only been with, you know, a girl I've been talking to for a while, and I feel like I have a connection to her. So, I think sexting responsibly has to do with not only knowing the person outside of social media, but having that type of sexual connection, I guess, even outside of social media. I mean, I wouldn't sext with a random girl on the street, but with my girlfriend, yeah. That's more responsible because I know my girlfriend. There's less of a chance for her to leak all my stuff out to the Internet and have it out there for everyone."

I ask if he worries that could happen anyway.

"No," Jeremy says. "I mean, I trust the girlfriends I've had. I obviously trusted them or I wouldn't have dated them, but I think, I find somebody who, even if we were to break up, it wouldn't be out of spite. I'd leave it on a good enough note where we wouldn't attack each other's personal things over social media. So, I've never really worried about it. Maybe I should be, but I never really worry about it too much now because I'm more responsible about it, sending it to people that I trust rather than people I just met over the social media site."

Jeremy doesn't worry too much about online privacy, either. He thinks about it a little bit, like everybody else, he imagines, but only a little. "But at the same time, I don't post anything that I should be worried about," he explains. "Not anything that could get me in serious trouble, at least. So, I mean, I worry about privacy and maybe my account getting hacked, but most of the time my account gets hacked and people just write stupid stuff, it's never really, like, they steal my information. So I'm never really too worried about it. I change my password probably, maybe every once in a while. Not too often, but I think I stay up on it enough where nothing serious will happen."

At the end of our interview, Jeremy says he wants to talk about freedom of expression. What he tells me reflects his generally positive attitude toward social media and new technologies, including sexting. "I think our generation has taken such a great step toward freedom of expression, and I think social media really amplifies that," he says. "There's a lot of diversity online. So I think when you go online, it gives you such a greater aspect and you can see the different views from people, and some people use it as a place for [judgment], but I don't think it's a place for judgment. I think it's a place for you just to express yourself and maybe find people with common interests. I think it really affects the world in more of a positive way than a negative way, if it's used correctly."

THINK (ABOUT YOUR FUTURE) BEFORE YOU SEXT

Jeremy has *some* company. About 14 percent of students in the online survey approved of sexting and see it as a positive, normal, and fun thing that people do. Some of these students spoke about how sexting can be great for a relationship, especially a long-distance one, or just wanted to say that there is nothing wrong with the practice. But the overwhelming message they sent was that you need to be responsible about it, which means not sexting people you don't really trust. And nearly half of the respondents were highly disapproving of sexting. Some of these students felt that sexting was "morally wrong" and even "disgusting," and "degrading," and a number of them were against the practice because they felt they got too many unwanted sexts. But the most common feeling wasn't that sexting is immoral, but that it's risky. These were students who spoke of sexting being *dangerous* because everybody knows (or should know) that photos and sexts can get passed around. These students believed that if you sext you're crazy because once you do, total disaster and ruin are just around the corner. People in this group expressed high levels of awareness that nothing is really private on the Internet and on phones, and also that everything you do on a device and that you send over the Internet is permanent. One sext can come back to haunt you even years from now.

This is one more sign of the heightened awareness that one wrong move, one risky post, one photo that you took just for fun could potentially rob you of your future hopes and dreams. These students thought that sexting is only "for people too immature to know any better," who haven't yet learned to "professionalize" everything they do online. Many college students would think that the idea of "sexting responsibly" is ludicrous. They would listen to Jeremy and shake their heads because, they think, he's kidding himself.

Those who turn to the technology and say that sexting is inevitable have a point. Smartphones have invaded nearly every aspect of our lives, so why should our sex lives be any different? Many students treat their smartphones as if they are attached to their bodies, and they have incredibly deep and complex feelings about their devices. Some can't bear to be without them, while others feel enslaved by them.

10

MY SMARTPHONE AND ME

A Love-Hate Relationship

As far as my phone, my wallet, and my keys, they have to be present. If they're not present, then I'm missing something, that's like me missing my heart or me missing my brain.

Jackson, senior,
public university in the Midwest

It's a perfect day because I left [my smartphone] at home and I feel naked right now, I really feel naked.

Katrina, senior, public university
in the Northeast

People expect you to be always online, or to respond as soon as you see something. But frankly, my dear, I don't give a damn. I'm a firm believer that people will write you whenever they damn well please, so there's no point counting minutes.

Omar, sophomore,
private-secular university

BLAIR: CLEARING MY MIND

"I just never really had a need for it, or even for a cell phone," says Blair. She's talking about her laid-back, almost ambivalent attitude toward social media. "Growing up, we weren't allowed to have [Nintendo] DS and PlayStation, but all my friends had it. I feel like me and my sister never really complained about not having it." Blair is so ambivalent about social media that she gave it up for Lent last year. "I was good without it, I could live without it, I feel like a lot of people couldn't live without it, and I feel like I would be one of the persons that would be able to live without it."

Blair is tall, beautiful, athletic, blond—the kind of sun-kissed girl you see walking down the beach carrying a surfboard in a teen movie. And Blair does surf, and sail, and is involved in "all water sports," she tells me. When I press her to say more about what prompted her to give up social media for Lent, I learn something interesting: it has to do with her smartphone. Or, more specifically, a particular incident that changed her relationship to the device.

"This summer I went without a cell phone for a week because I dropped it in the ocean," she says with a laugh. "I was working, and I'm a lifeguard. So you sit for half an hour, and then you're on the stand, and then you're half an hour off, and everyone's just always on their phone. It's boring after a while [being on your phone]." When the phone fell in the water, Blair was a little traumatized. But then she began to feel better. And better. "I did so much more that day, just, other things that I normally wouldn't have done if I didn't have my phone. And my mind felt *so* clear after, like, one day without it, and it was such an *amazing* feeling that you're not just on your phone the whole time, in your face. And it was relaxing, too. You just don't have to be in touch with everyone 24/7."

Between the incident this summer and Blair's Lenten sacrifice, she learned that without social media and her smartphone, she was able to concentrate more, to live in the moment. Even while she was sailing this summer, she noticed how everyone was busy getting notifications from their friends all the time and not really paying attention to what they were doing.

Blair's relationship to her cell phone has changed since she dropped it into the water. She has a new phone, but she often doesn't have it turned on. And if it is on, she'll put it in a mode that only allows her to receive phone calls—not even text messages. "I actually found that I don't care if I get a text message, or if I respond right away," she says. "I keep leaving it on silent and not knowing if I get a text. When I have time I'll look at it to see if I have one, but I'm not constantly checking. I don't really care if I keep on getting text messages. Text messages will be there all the time. If a phone rings, then I can pick it up, because a phone call I feel like is more important than just a text message." And she no longer has her social media apps on the home page of her phone. Instead, they are hidden in a folder "a few pages back" to make them harder to get to.

Blair has made an effort, of late, to not let her smartphone take over her life. She sees people all around her who are obsessed with the battery life on theirs. "Everyone's like, 'Oh, my God, it's on, like, 5 percent, oh my gosh!' I actually turned it off recently, and it's so much better to have one less worry I can't do anything about it." If Blair is aware that her battery is low, it stresses her out, but what she doesn't know alleviates that unnecessary stress.

Blair would prefer that smartphones didn't exist—or, at least, if she didn't have to contend with them. "I wish I grew up in a different generation, just for that reason," she says. "No one makes eye contact anymore. I'm always smiling, being superfriendly, and no one does that anymore because they're always on their phone, [with their phone] in their face." Blair's friends are always on their phones, even when they're out to dinner, and this upsets her. "I'm like, 'Oh, phone party,'" she says with a sigh. "Everyone's on their phone, and I'm never that person that's on their phone. If I'm spending time with people, I want to spend time with *them*, not with some person in cyberspace." It frustrates Blair to no end that her friends won't put away their phones while they're socializing. "I'm like, 'Guys, phone party? Really? *Again*?' Then they'll put it away a little bit, but still, I can't stand it, so I don't do it. Like, I try not to be that person."

THE JOY OF BEING UNPLUGGED

My first visit to a campus for this project happened to be shortly after spring break at a Catholic university. Many students had either been to places or in situations where they'd had to set aside their smartphones for the week. And they couldn't stop talking about how much they *loved* their life without a smartphone—at least for a bit.

Like Blair, it frustrates Gina to no end when she's in a group of friends and everyone is on their phones. She's also done it herself and then regretted it. Suddenly everyone around her will laugh, and when she looks up from her phone, she realizes she missed whatever was so funny. "I think technology definitely pulls us away from the real world, which is okay at some times, but I think most people are not as engaged as they should or maybe would want to be with the technology that we're always using." Gina is guilty of texting and using her phone in class, because it's so hard to put it away. She brings her phone everywhere, and even in the cafeteria she'll have it out on the table. She's a little embarrassed as she tells me this.

"I think what's sad is I justify it with myself and other people be-cause with school, we're constantly getting emails, and if you miss one email, that could be something really big that you're missing, like a group meeting or someone who needs you to, like, for volleyball, sign a waiver for something or picking up graduation gowns," Gina explains. "It's just, we have so many things that we need to keep straight, so we're constantly getting emails. We're constantly looking at our calendars, constantly trying to figure out times to meet with people. So you almost need to have your phone or you could be missing out on something important."

Almost everyone Gina knows expects her to be constantly available. I heard this same complaint from one student after another. But then Gina mentions her recent time away from her phone.

"I was just away for spring break," she says. "Didn't touch my phone for the whole week just because we could use Wi-Fi, but it was only two hours a day, and I was like, 'Whatever, I'll just let it go.' It was *great*. Like, so liberating. It didn't feel like I needed to talk to *anyone*. I didn't have to like talk to someone, and time wasn't really a factor of my day, I'd do whatever, whenever. I didn't have to worry about anything." Gina paid for all this freedom afterward, however. When she came home from spring break and turned on her phone again, it was so full of emails, texts, and Facebook notifications that she was overwhelmed and incredibly stressed out. "But during it, it was *awesome*," she adds, laughing. It made Gina "feel more on vacation," and she loved how everything was more relaxed—how everything seemed less scheduled. Gina spoke of her smartphone as a kind of taskmaster, controlling her life and keeping her in line, always reminding her of her endless to-do lists.

If being without a smartphone was really so nice, I ask Gina if there's any reason she can't adopt a phone sabbath of sorts.

"I do wish I could, but I feel that I just can't," Gina tells me. People always need her—her teammates, her coaches, her roommates, her professors, her parents. "If I don't have my phone and I'm not available, I feel like then if they need something from me, it's going to reflect poorly upon me that I'm not responding quickly. Because that's another thing: I want to respond to people quickly because if I reach out to someone, I would hope they would like respond in a timely manner."

Kristin, a junior at the same university, had a similar spring break experience. She, too, starts out by telling me how she's on her smartphone all the time and also how she "doesn't know how she existed without a smartphone" before they were invented. But then spring break came along.

"I went on a cruise, and I turned my phone off for the whole week and it was great! I didn't have that fear of missing out. I didn't want to turn it on," she says, laughing. "When we got back, I was like, 'I wish I could just keep it off forever.'" I ask Kristin to tell me why she feels this way. She explains, "I think that being in constant contact with the world sometimes is overwhelming, so if I can get my email at all times of the day, it's like, I always have to check my email, you know? . . . .It would be nice just to be able to relax and not have the expectation that I'll be getting people's messages and having to sometimes act on them. So it's nice just to be unreachable for a little bit."

I ask Kristin whether, given how much she loved being away from her smartphone, she's considered making herself "unreachable" occasionally. "Sometimes I do," she admits. It's okay, she thinks, to set aside her smartphone when she is studying for an exam. On Sundays, too, she'll just "put it in another room while [she] goes about her business." But setting it aside in the long term seems infeasible. "I don't think overall I would get rid of my smart phone because it does have its uses, and I think it's hard to function in society if I were to completely make myself unreachable."

In addition to students like Gina and Kristin, I heard from others who'd gone on volunteer trips to far-flung locales and who had no choice but to give up their smartphones for the week. They waxed poetic about what it was like to socialize when no one was looking at their phones.

When I met Amy, she had just given up her phone—reluctantly, and at the urging of her mother—while on a trip. In the end, she was glad for the experience. Amy spoke of never being without her smartphone, of how she worries about not being immediately available, how the expectation these days is 24/7 accessibility. But then her mother complained that Amy's face was "buried in [her] phone all day." So her mother made Amy leave her phone on the kitchen counter, told her that for the entirety of spring break she needed to disconnect, that she wasn't allowed to bring her phone up to bed with her, or carry it around at all, not when in the house and not when she went out either. "And it was awesome, you know?" Amy says, upon reflection. "And [my mother] was kind of policing me about it, because she would see me go and check. But after a day or two it didn't feel like I was missing out on anything. I was talking to my family more, and I was watching TV and actually watching it, not like, half watching it and half scrolling on Reddit. And it was a really, really nice change. And every so often when I'm home, I do that same thing without [my mother] telling me to. I just leave it on the counter, and I go about my life." "But, she adds, "I don't do that when I'm at school."

This was a familiar refrain from students: on vacation, sure, on spring break, sure, on a service trip, sure—I can set aside my smartphone. At school, that's impossible. Most students feel pressure to be available seven days a week, twenty-four hours a day. But if they felt they had a legitimate excuse to sign off for a time, they did so happily.

One student, Avery, had spent the previous semester in Africa, and the entirety of it unplugged. I ask Avery how it felt to go for that long without being connected.

"It was just really good," she says. "I felt like it was a legitimate excuse for me [to unplug] . . . .because I don't like to be constantly plugged in to social media on my smartphone or constantly texting people back. I felt like not having Internet and not having Wi-Fi was a legitimate excuse for me not to keep up and replying to this, commenting on that. So it was very refreshing." Avery especially liked how living unplugged changed her relationships with others, especially those in her study-abroad program, and she enjoyed her activities more because she wasn't on her phone all the time. "We went on camping trips and safaris and awesome beach trips. I cannot image being on one of those and having us all staring down at our phones. That just sounds horrible." Avery bemoans the fact that Wi-Fi will soon be everywhere, all over the world, so being un-plugged will not be easy.

I ask if it's been strange being back and being connected again.

Avery says the hardest part is meeting the expectations of others. "People weren't expecting responses extremely quickly when I was there, and now they are, because they know I'm back. But I still don't give them very prompt replies." Avery's time abroad has changed her relationship to her smartphone. She leaves her phone in her room more often now, if she goes to the cafeteria or even on a weekend trip. She finds her smartphone distracting, and she doesn't want to be like her friends who are "totally debilitated without their phone." Avery doesn't like the way people com-pulsively check their phones or pull them out if there is a quiet or awk-ward moment. When she walks across campus, she sometimes counts how many people are actually not looking at their phones, and is dis-mayed to find how few are paying attention to the world around them. She guesses maybe only one in five is looking up. A smartphone "takes you out of the moment, and it brings you to another place that is probably not important at all," Avery says.

There were other students like Avery, who told me of trying to give up their smartphones periodically each week in order to study. In the online survey, of the students who answered the essay question on this subject, 70 percent said they intentionally take breaks from their phones.

I heard from students who played tricks on themselves, like inten-tionally leaving their chargers at home, and making a pact that, once the battery ran out, this was it: no more smartphone that day. Some students take simple, short breaks, leaving their phones at home during social situ-ations or dinners, putting them in another room while studying, leav-ing them home for the day on purpose. Many students spoke of that corner, way down in the third floor basement of the library, where the Wi-Fi doesn't work and where, for exactly this reason, they—and so many others—cram in with everyone else to study. Many claim they simply can't study or concentrate unless they unplug from their phones.[1] And many talked of a daily battle to stay away from their smartphones while social-izing and studying, doing things like turning their smartphones to "air-plane mode" and putting them face down or even relinquishing them to a neighbor in their residence halls when they had an exam coming up. It is indeed a battle, too, and the students are fighting hard, to the point where some of them sound like they need a kind of Smartphones Anonymous support group to help them work through and deal with their addiction. I even met a few students who told me that one of the reasons they liked going to church—and made a concerted effort to do so—was because it gave them a reason to get away from their phones for at least one hour a week. Students would nearly always mention how hard it is to be away from their phones at first, but most ultimately find it liberating.

Other students said they take far longer breaks and specifically look forward to circumstances where they will simply not have access to their phones for days or even weeks at a time. These students are exuberant about the time they spend off of their phones.

Mercedes is one such student. She sought relief from the pressure to be constantly available. "I feel like you can be reached at all times, all day, every day, via many different outlets," she complains. "That can be difficult, especially because [people] know you're on your phone, they know you have it with you, so why aren't you answering? So you should probably always respond immediately." Mercedes goes to Catholic Mass weekly, and she never uses her phone during church. She enjoys and looks forward to that hour away. When I ask Mercedes whether she ever feels tempted to take a peek at it during the service, she tells me, emphatically, "Absolutely not." "I usually leave it in the car or make sure I turn it off," she says. But then she goes on to explain that the first thing she does when she gets back in the car is check her phone to see who's texted or what's happened on Facebook. "I mean, I'm sure the first thing I'll do when I leave this interview is I'm going to check my phone. That's definitely going to happen."

At a Catholic university in the Northeast—the same one where I met Blair—I happened to be on campus a couple of weeks after students from the psychology department had done an experiment in which they required everyone to put their smartphones into a basket for the duration of dinner in the cafeteria. Everyone was talking about it.

Among the many students who mentioned this experiment was Emily, who'd gotten off Facebook and Twitter for the summer, and who told me she always turns off her smartphone when she goes on vacation. She doesn't like the compulsions that both social media and her smartphone make her feel—she can't resist the urge to "click it," she says, so the best option is to just "get rid of it for a while." When she does this,

Emily feels like "a burden is released." Like most everyone else, Emily also detests how people are on their phones at the dinner table—though like most of these same students, she is guilty of doing it. This is when she tells me of the psych department's experiment. "They put little bins on all the tables, and you had to put your phone in it, face down," she explains. "You couldn't check it all dinner, and everyone was actually having a conversation together. Which is pretty rare." Another young man who mentioned the experiment told me how, at some point in the middle of dinner, with everyone's smartphones sitting untouched in the baskets, he realized that the entire cafeteria was "louder" than normal—because people were talking and laughing so much more than when everyone is on their phones.

Of the students who answered the essay question about whether they take breaks from their smartphones, about 30 percent said they never do—at least not willingly or intentionally—and a fair number of them expressed a kind of superiority about this. They felt a sense of *pride* in being able to have their smartphones with them at all times, while not feeling addicted, distracted, or dependent on their presence. In these answers, students expressed judgment about those who "lacked the self-control" to stay off of their smartphones without needing to create circumstances in which they can't use it. Students who have their smartphones with them at all times and have the will to resist feel smug. They are well aware that they are an unusual breed.

But almost every student I spoke with complained about how smartphones are detrimental to social life. One young woman called them the "new yawn," because yawns are supposedly contagious, and once one person pulls out his or her smartphone, everyone else follows. I heard again and again about deals students made with each other to try to prevent this from happening during dinner, with everyone agreeing to relinquish their smartphone either into some sort of bag or at the center of the table, with the first person to give in and look at his or her phone having to leave the tip or pick up the check.

Besides creating a pressure to be "constantly on," the most common complaint I heard about smartphones was their disruptive presence in social situations. Despite this resounding frustration, the students who complained also tended to admit (rather sheepishly) that they were guilty of such behavior themselves. And despite the rather notable downsides of smartphones, it was rare to encounter a student who wanted to do away with them altogether, or who would be willing to give them up even if this sacrifice came with great social satisfaction and the relief of not being constantly available, day and night.

ALWAYS "ON CALL"

It used to be that certain doctors, police officers, firefighters, and, well, the president were the only ones who needed to make themselves available at all times, even in the middle of the night. But according to the students who chose to answer the optional essay question about this very subject (Due to the prevalence of smartphones, do you believe we are now expected to be available 24/7?), today, we are all like doctors. At least we act that way. College students believe that smartphones have created an expectation that they are "on" all the time, day and night.[2]

Sixty-nine percent of the students who answered this question did so in the affirmative. And mostly, they hated the expectation. A few students seemed to thrive on the pressure to be available at all times, and they tacked on phrases such as "And I love it!" or "I like it that way" or "And I think it's great!" to their answers. But these students didn't have much company.

Most students used an extremely negative adjective to describe the perceived expectation of constant availability: stressful, awful, terrible, frustrating, impossible, exhausting, unrealistic, absurd, unfair, ridiculous, devastating, and unhealthy. People expect you to answer them "even while I'm asleep," one student said, while another lamented, "If I don't respond immediately I need to explain myself *now*." A number of students commented on how, because of this expectation, it's never been more important to set boundaries around personal time, and several wrote that despite this expectation, they simply don't make themselves available. One student who thought this expectation was ridiculous and overwhelming actually added, "But it seems I am alone in feeling this way," which nearly made me laugh out loud because nothing could be further from the truth.

Even the 25 percent of students who said they did not feel they were expected to be available all the time were affected by the expectation of this perceived by others. Some were defiant, saying, "I don't allow myself to be" or "I draw boundaries." But just over a quarter of those who answered no took the question very literally: people can't expect you to be available when you are *asleep*—the implication being that during waking hours you *are* expected to be available at all times.

It's no wonder that so many students feel such relief when they unplug.

But not everyone longs to unplug even for a little while—not for themselves or even for their friends. In fact, several students I met seemed to live for their smartphones and feel that they, quite literally, could not live without them.

CHERESE AGAIN: YOUR SMARTPHONE OR YOUR LIFE

"I like my phone *a lot*," Cherese says, by way of beginning what will become a very long, rather amusing, and somewhat shocking conversation about her relationship to her smartphone. Cherese tells me she's had a smartphone ever since they came into existence, and before that her phone had one of those slide keyboards with buttons on it, which she also loved. Her little sister broke that phone one Christmas Eve, and Cherese tells me that she was "just getting ready to kill people because I needed my phone." Cherese had to wait until the day after Christmas to get another one, so "it was just not the best of days." Her phone is her "safety net," she says. Her security blanket.

As with so many people, Cherese uses her phone for *everything*. Her calendar is on it, she docs her banking on it. But unlike other people who love their phones, Cherese once risked her life to save hers.

"So this summer, I was robbed," Cherese tells me. "It was when I was on the bus, and I actually paid somebody just to give me my phone back, because my phone was just, like, *that* important. It was a group of them. . . . I paid them a hundred and fifty dollars for the phone."

I have a number of friends who've been mugged for their smartphones—but everyone I know has simply handed over the phone, to avoid getting shot or stabbed in the process. "So these people mugged you for the phone, and you ended up giving them money?" I ask, to make sure I understood Cherese correctly.

"Yeah," she confirms. "And they had already stolen my wallet, but I was just really concerned about the phone because I can go and cancel the cards. [Afterward] my parents were telling me, 'That wasn't the best idea to get off the bus and pursue it,' but I had so much stuff on my phone."

"So you actually got off the bus to pursue them?" I ask, still incredulous.

"Yeah, I followed them to get the phone, and then the police were like, 'Oh, that wasn't the best idea,' and I was like, 'Well, I've lost my phone before and I really need my phone.'" When I press Cherese to tell me if the phone was truly more important to her than her own safety, she laughs. "At the moment, I really thought it was. That's the only thing I could think about. I wasn't thinking about my wallet, my keys, or anything like that. I was just really focused on the phone because it was just like, 'I have everything on here. I really can't survive without my phone. I really need to get my phone right now!'"

Earlier on, Cherese had told me that she once went on a seventeen-day "fast" from social media. I ask her now if she'd ever done the same thing with her phone. No, she tells me. She wouldn't even attempt such a thing because she wouldn't be able to make it.

"Like right now," she says, "I left my phone in the other office. I'm just thinking like, 'Wow, what if [the woman working in the office] leaves with my phone sitting there?'" Has she been stressed about it the entire time we've been talking? "Uh, a little bit," she admits. "It's just like, it's my *phone*. When I leave the house, I don't worry about my keys. I worry about my phone. So, I'm just really concerned."

"So you'd rather be locked out of your house?" I ask her.

"Yes," she says simply.

I tell Cherese that we're almost done with our conversation, and that she'll be able to retrieve her phone soon.

"Oh, that's okay," she reassures me. "I don't think that [the woman is] going to leave because I can still hear over there." Cherese goes on to tell me that she's been monitoring what was happening in the office next door the whole time we've been talking to make sure that her phone was okay. The reason she left it there in the first place is because the battery dies quickly, so the all-important charger is there too.

Just seeing her phone sitting on the table makes her feel more reassured, Cherese tells me, especially in social situations. The presence of her smartphone makes her feel calmer, and it serves as a helpful heads-up to others that she's always available to interact with them. "My phone affects my happiness. . . . Having my phone and knowing that it's here if I need to get ready to use it, it's just reassuring because I know that I kind of won't have to talk to other people. Sometimes I'll just use my phone as a way to not talk to people. So, it's like, 'Don't you see that I'm on my phone, so don't say anything to me.' "

LIFE BEFORE SMARTPHONES VERSUS LIFE AFTER SMARTPHONES

Cherese's relationship to her smartphone is unique. She is certainly the only person I met who was willing to risk her life for it (or at least, the only one who openly admitted this). More common were the students who brought their smartphones with them everywhere, who were constantly checking them, who felt that smartphones brought a certain amount of really great convenience, and at the same time, a certain amount of really unfortunate responsibility. It was rare for students to speak only favorably about their smartphones, and even the ones who really liked their phones could still appreciate a forced vacation from them now and then.

But even some of the same students who found great relief in having been forced by circumstances to give up their smartphones for a time sat with their phones in their laps for the entirety of our interview, cradling them, turning them around and around in their hands, though not checking them directly. It was rare that a student actually checked a phone during our conversation—it happened only a couple of times. And even though so many students longed for the opportunity to be unavailable for a while, to not have to be constantly "on," there were few students who said they'd like to give up their phones entirely. There was a clear demarcation for them between life *before* and *after* the smartphone.

"I remember my life before the smartphone," says Matthew, whom we met earlier. "I had a flip phone, and I had to hit the same button, like, four times to get to the right letter and it was terrible, and now, everything's so easy. My phone guesses what I'm trying to type. It's really nice. I can just talk to my phone and it'll type it for me. I don't have to actually even have my hands on my phone when I'm in my car, to use it." Matthew goes on and on about the wonders of this, and how it's also made get-ting onto social media easier as well. But then, once Matthew finishes listing the many conveniences smartphones offer, he begins to mention the drawbacks. "But also, I feel like people just put it up to their face a lot of the time, and then there's a lot of stuff that's happening that they're missing. Like, at lunch, I'll pull up to a table with six people, and we'll be eating, and five people will be on their phones and it's just me and this one other guy talking." Like so many others, Matthew is dismayed by the negative consequences of smartphones in social situations, especially if he's out on a date. "If I'm on a date and my date pulls out her phone in the middle of me talking or something, I'm like, 'I don't want to date you at all now! We're done! Like, put your phone away, I'm right here,'" he says, laughing.

While this is a deal-breaker for Matthew when he is on a date, he finds it incredibly difficult to put away his own phone. "I feel like I, I *can't*," he says with a groan. "Especially lately. I dropped my phone, and it's been weird, so my battery hasn't been lasting long at all. I literally have to rush between classes to go charge my phone so I can know what's hap-pening. At any point there could be an email that is important for what I'm doing later, so I get real anxious if my phone's dead or it's not on me." Matthew gives me an example. The previous night, he had been trying to text his mother and his girlfriend. His phone was dying, but his charger was across campus. "I could *not* concentrate, and I think after ten minutes of trying to fight it, I ended up walking all the way across campus to get my charger and walking back so I could plug it in to the wall." He seems embarrassed and adds, "Yeah, it is *bad*." I mention that it sounds like his life revolves around plugging in his phone, and he admits that before our interview his phone died again, so charging it is the first thing he's going to do when we're done. Then I ask if this stresses him out—that his phone isn't currently charged. "I am a little stressed out," he says. I tell him not to worry, we're almost done. "Well, before my next class I'm going go charge it, so everything will be okay."

Matthew and his phone would be reunited soon.

He wasn't the only student I needed to reassure about this. One of the things I had to do quite often during interviews was tell students not to worry—the interview was coming to a close in just a few minutes, so very soon they could return to their smartphones.

I heard from many other students about battery stresses, too—how, with smartphones, they don't last long, so it's important to carry a charger around with you. Some students asked to charge their phone in the room where we did our interview. I met one young woman in the Midwest who calls her smartphone "Meg"; she wasn't the only one to give her phone a name. It was common for students to use phrases like "my smartphone and I are very close" or "my smartphone and I are always together," as though their phone was another person to whom they felt great attachment, almost like a significant other. Some students laughed and said things like, "We have a complicated relationship." They both loved and hated this object they carried with them everywhere, for better or worse.[3]

The relational language the students used, the "we" and "my smartphone and me," the tendency to personify smartphones in some way all make it seem as though smartphones are not merely objects to us—they are like people with whom we develop relationships. We love them, even cherish them, then resent or even hate the things they "demand" from us, like 24/7 availability. Smartphones can be as needy as an insecure boyfriend or girlfriend who constantly seeks reassurance that we're still there, that we care.

One student I interviewed claims that our relationship with our smartphones is like a romance—it's really intense at first, and then you get more in control of things as you get used to it. "You sort of have to get the handle of it and master the way that you use it because it can be overbearing," he says. "But when used properly, it's a good thing."

Then there are students like Jackson, who likens leaving his smartphone at home to going out without his brain—a comment that makes both of us laugh. But the thing is, Jackson is serious, and we both know it. "Everywhere I go, my phone goes," Jackson says. "My phone's kind of old, so it tends to die, so I take my charger too, but yeah, everywhere I go, the phone has to go with me." Jackson is the unusual student who never tries to get away from his phone, never wants a break from it, and seems untroubled by the constancy of its presence in his life and the way it compels him to be on it all the time. He doesn't find his phone to be a distraction from what's essential. His phone *is* essential. "As far as my phone, my wallet, and my keys, they have to be present. If they're not present, then I'm missing something, that's like me missing my heart or me missing my brain." Jackson is smiling, laughing, and nodding at me as he says this.

"Those are the three essential things that get me through my day, I guess you could say. If I have my phone, I'm able to contact people; if I have my wallet, I can, you know, survive as far as finances; and my keys, of course, get me in my house and my car."

Like Jackson, Daphne lives for her smartphone. She has her phone in her pocket for our interview, she tells me. And then she goes on to offer a comprehensive list of every phone she's ever owned, starting with her "first flip phone" and including five other "craze" phones, all the way up to the latest iPhone. It's really important to Daphne that she has the most cutting-edge phone out there. She's constantly on apps, or texting and chatting with people. "I always have the thing. Something might happen, someone needs to talk to me, or my family needs to get ahold of me. . . . But then again I'm on my phone constantly texting my two close friends and my boyfriend all the time. And my grandma even texts me too because we just got her a smartphone because she wanted one."

I ask Daphne whether she feels that smartphones make it so that she has to be available all the time—that constant complaint I hear from her peers. No, it turns out. "I just feel like I'm going to miss out on something," she says. "I download the apps just 'cause it's there and just 'cause I got it. So when I'm bored waiting for a class or I'm looking through my phone, creeping people, seeing what's going on, killing time." Daphne insists that she's not one of those people so enslaved to their phones that they would "jump in front of trains to go get their phones," though. "Which I would never do," she says, "because I'd just buy a new one. Because it's easily replaceable. . . . And I'm like, there are people who jumped in front of a moving train to go get their phone! Which I don't think is good, but it's just, everyone is so attached and constantly needs to be talking to people and showing the world what they're doing. Which is silly, but then again, I do it as well."

Daphne does try to draw the line and not check her phone in social situations, especially when she's with her boyfriend. She has stayed off her phone on vacation (her mother made her) and during a church retreat, and after a while "it feels good when it's not there and you have nothing to worry about," she says. "But then again you feel like you need it." Daphne expresses a deep desire to be with her smartphone at all times. It's a security blanket and seems to give her emotional support.

"It's just that I feel empty when I don't have my phone," she tells me.

Stacie, a junior, speaks at first about how much she loves going on camping trips where there is "zero service, so there's zero communication." She "loves it and it's so stress-free." Stacie thinks that there is "something really relaxing" about not having your smartphone with you, and that "you become a lot more self-aware without your phone."

Despite this, not only would Stacie never want to go back to a time when smartphones didn't exist, but she believes we simply can't. We're too dependent on them, even for basic things like getting around. "People don't know how to use maps anymore, they don't know how to use dictionaries, and I think that if you took a person to the next state and put them on a corner and then just left them, they would need a phone, they would absolutely need a phone," she says. "Everything is so easy now that we don't really have to be independent, you know? We're dependent on our phones." If smartphones disappeared, Stacie thinks "we'd all be lost."

Stacie sometimes tries to leave her phone in the next room, so she's not always checking it. But both she and her friends go crazy if they try to get in touch with someone and don't get an immediate answer. They get mad, and they panic. Even though taking a break from your smartphone can be "relaxing," Stacie also claims "it's kind of scary, you know, not having that crutch," the feeling of holding a smartphone in your hand and knowing it's there with you. "I think you can absolutely just leave your phone at home, but is it safe?" Stacie wonders. "That's the real question, you know? It's one thing to be like, 'Oh, I don't want to communicate with the world, I just want to be at peace and with myself,' but it's no longer *safe*, especially for a girl my age. It's definitely not safe to just go out without a phone and do whatever you want. So, you know, you really can't be without communication."

That comment about safety—that it's "not safe" to go about anywhere without a phone—came up again and again, nearly compulsively so, as did the one about how we "must" be with our phones at all times because you never know when there might be an emergency and someone needs to reach you or you need to reach someone else. Even though entire generations grew up without such "safety measures," students lean on this notion because it makes them powerless to put their phones down. Being without a phone just isn't practical, since you never know when tragedy might strike.

UNPLUGGING ... MAYBE FOREVER?

Occasionally, I did meet a student who loathed what smartphones have done to our lives and fantasized about going back to a time before smartphones existed or forward to a time when they would no longer be needed. These students regarded smartphones as a kind of necessary evil.

When Marcus walks into the interview room, the first thing I notice is the colorful black eye on his face. He's tall, muscular, with short, curly hair and has the look of a tough guy. But as soon as we start to talk, it's clear that he's sweet and sincere—the black eye is from playing football. He's very engaged in his Catholic campus's life, too. In addition to being on the football team, he's heavily involved in his business major.

Marcus cannot stop talking about his smartphone. It starts off as a love story.

"I don't want to sound like I'm in love with it, but you know, it has my agenda on it, it has my work schedule," he says. "Honestly, probably texting is the least I do on it. I use my phone, you know, [for] pictures, my music's on there, ...so I use my phone a lot. I charge it, like, one and a half times a day, so it's definitely a necessity, I would say, in my life." Marcus, like everyone else, feels he needs to be accessible 24/7 and tells me so.

But the love story between Marcus and his smartphone quickly takes a turn for the worse. "It's a shame," he says, that having a smartphone makes him feel like he needs to be accessible all the time. "But, you know, on my bucket list, I have 'retire,' and then I have 'get rid of my cell phone.' You know: get a home phone." I am surprised that among his dreams would be getting rid of his smartphone. I ask him to explain why.

"I did not want a cell phone," he says emphatically. "I think it controls everybody's life. I just, I *don't want one*, but you know, I *have* to have one." Marcus believes that it is impossible to survive modern life without one, especially as a college student; smartphones are practically a requirement—even if having one makes you feel unhealthy as a person and damages your relationships. "People are using electronics to communicate, and it's efficient for me, so it's working to my advantage," he says. "But I would love a life where I just didn't need a cell phone. Where if I told somebody I was going to be at a meeting in a week, at 2:00 p.m. in this room, I would be there and so would they. You know, I wouldn't need to keep texting them and reminding them. That's a world I would like to see, but I don't think that's happening," he adds with a chuckle.

When Marcus goes away with his family or friends, he leaves his smartphone behind. But he definitely believes that, for business purposes, he'll need to carry his smartphone at all times. It's unrealistic to think that he won't. "I'm a marketing minor, so social media is how you get things out," he explains. "You can get a thousand people to join a Facebook event and you know that can boost your sales, boost revenue, boost profit, and that's the world I'm in right now." But talk of how much Marcus will need to use social media and his smartphone for his career leads him right back to talk of retiring. "You know, I'd prefer to live my life without it, but it's not going happen," he says, a bit sadly. "I've come to the realization that, when I retire, then I'm going to get rid of it. *That* will happen. It's *going* to happen."

When I travel a little farther north, to a different Catholic university, I meet Stephanie, a senior who has a smartphone but *hates* it. In fact, Stephanie's exact words are "I hate the thing," which tells me right away that she isn't the type to personify her smartphone. "I mean, I'm connected to it," she admits. "If I can't find it, I freak out a little bit." Part of Stephanie's hate stems from her worries about the object itself: Will she lose it? Will it break? What if she forgets it somewhere? Smartphones are expensive, and Stephanie doesn't want to be out all the money she spent to buy hers. She does find it convenient, though. "It beats carrying around a laptop," she says, and she thinks it's helpful for staying connected to her mom. "And, you know, if she forgets to feed the dog she can tell me," Stephanie says.

Stephanie rarely takes a vacation from her smartphone, but "it feels good" when she does. "It's liberating, you know? You're not clinging to this thing that's in your pocket or your purse and you can just do something for yourself. Like, you don't have to worry about someone texting you or missing a call. Or even going on social media. You can just go out and just appreciate the world," she adds, chuckling. Stephanie notices a "huge difference" in her mental state when she has her smartphone on her as opposed to when she doesn't. When she has it with her, she worries about what everyone else is doing, but if she leaves it home, she thinks about what *she* wants to do instead. What's worse, smartphones make us more dependent on social media, in Stephanie's opinion, and "being dependent on social media means losing a bit of yourself" because you become so focused on everyone else. "So, getting rid of [your smartphone] helps you reflect on yourself. And so being attached to [your smartphone], making it a part of yourself, takes away from you, if that makes sense? It's like, a *leech* or something."

Stephanie—true to her philosophy-major self—takes time to imagine what her favorite philosopher, Heraclitus, would say about social media and smartphones. "Heraclitus was very much about change, and he says change is a good thing, seeking knowledge is always a good thing, also self-discovery," she explains. "People are constantly changing, the Internet's constantly changing, he would accept that. However, the truth part of it, finding yourself, you're distracted by that with social media. Social media is distracting you from finding your true self. I think that's what he would think." Stephanie pauses to reflect a little more broadly. "I feel like the world is losing its self-identity," she says. "You know, you have people looking down so much at their phones they don't get the chance to look up and see the world around them. And social media can also prevent you from meeting someone, who might end up being very important later. You're just so tied to this little device in your hand, and I think that's really unfortunate. I wish people could just break this habit, but habits die hard. The way our generation is going, it doesn't seem like it's going to change ... and, I mean, it's unfortunate."

SMARTPHONES ARE HERE TO STAY—AND THEY ARE EVERYWHERE

Many college students resent the ways in which smartphones are changing people's expectations, and many of them long to take breaks, even if it's difficult to do so. But students love their smartphones, too, and the devices are ubiquitous among the college population. Only 4 percent of the students I interviewed, and 5 percent who answered this question in the online survey, did not have smartphones.[4]

For the most part, the students who do have smartphones always have them on hand, and many students draw comfort from their presence—a kind of reassurance that all is well.

"The act of having my phone in my hand probably brings some kind of comfort to me, like, that's my armchair," one young man explained. He used to think this was kind of a negative thing, to be so dependent on his phone, and would try to convince himself to stay off of it, but lately he's changed his mind. As he told me, "If my phone is my armchair, I don't think it's that bad. It's not bad *enough*. The benefits outweigh the drawbacks."

This seems to be the prevailing feeling. College students may be frustrated and even dismayed by how smartphones have changed their lives and especially how they can affect—negatively—social situations, but almost no one is ready to give them up. Some students fantasize about going back to a flip phone or look forward to someday "retiring" from their smartphone, and students may indeed take breaks and engage in negotiations around the extent of their smartphone usage and their availability, but they are too accustomed to the conveniences of smartphones to actually live without them.

In *Reclaiming Conversation*, Sherry Turkle also writes of the ways in which we both lament the changes technology has wrought and at the same time "resist" changing our relationships to it. She, too, heard interviewees in a workplace speak about their smartphones being problematic on many levels, but when she asks "why they continue to bring their devices to meetings, they say, 'For emergencies.'"[5]

There is no doubt that students love the convenience smartphones offer them, the easy access to just about anything they want, the constant availability of the Internet, the GPS function, how smartphones make everything "immediate," that smartphones make it easy for them to keep track of their busy lives and commitments, that they can play games and listen to their music on them. A few students made simple effusive comments about their smartphones, such as "It is my life," "I love it—all of it," "It's like having the entire world in the palm of my hand." One student went so far as to say, "It makes me feel like I can do anything!" There were a number, though, who felt that even these positives were also negatives, and that smartphones were ultimately "a curse in disguise," in the words of one; others, when asked what they liked most about their smartphones, replied, "Very little."

There is a prevailing sense that if someone just did away with the things, all of us could get on with our lives in happier and better ways.

The *burden* we are carrying around because of our phones would be lifted if they would only disappear off the face of this earth. These tiny, light, pretty, shiny devices have come to represent an outsized weight upon our shoulders—we look at them and see our to-do lists, our responsibilities, other people's needs, our perpetual inability to keep up, the ways in which others constantly judge us, everyone's successes amid all our failures, among so many other stresses—stresses that feel more like thousands of pounds than a few ounces. At the same time, we see them as our escape from boredom and loneliness, our connection to loved ones and friends, our guide when we are lost, the archive of our best hair days and most memorable moments, the diaries where we place all our most intimate feelings, hopes, and dreams.

It is clear that students' relationships with their smartphones are intense and often fraught—much like the relationships we have with other people. Indeed, students are at least as attached to their phones as to their friends. And while they might sometimes wish they had never met their smartphones, ultimately they can't give them up.

Or can they?

11

TAKING A TIMEOUT FROM THE TIMELINE

STUDENTS WHO QUIT SOCIAL MEDIA AND WHY

I went to bed on Instagram, I woke up on Instagram, and I most likely went to the bathroom looking at my Instagram. That's why it had to stop. No object or person should have that type of stronghold on you.

Zooey, first-year, evangelical Christian college

I have deactivated my Facebook account several times just for temporary reasons. If I start becoming too hard on myself and thinking that I need to be doing what everyone else is doing or having more fun or traveling more or just any number of things, than I delete the account and make myself take a break. The first couple of days are hard to not just log back in and check what's going on, but the more time I spend away from Facebook, the easier it got, and once I even got to the point where I no longer thought about checking Facebook. Eventually I reactivated

the account because it was summer break and I was
bored. But I did find that I was happier without
Facebook and I hope to delete it completely soon.
I think it is an unhealthy obsession for me and I'd be
better off without it in the long run.

Tamara, senior, public university

LAUREN: I'LL QUIT SOON

Lauren, of the dating fail, is carrying eighteen credit hours this semester and also has a part-time job. She's a sophomore at a Midwestern public university, and in a perfect world, she might try to join a sorority next year, but who knows if she'll have the time? As Lauren begins to answer my questions, tell me about her life, and share her opinions, I'm laughing nearly constantly. She's hilarious—by far the funniest student I interviewed.

What I learn quickly, loudly, and clearly is that Lauren *hates* social media, *hates* smartphones. She's *on* social media, she *has* a smartphone, she just loathes them. First of all, she thinks people lie constantly online. "There are millions of people that lie on social media daily," she says. "There *could* be authentic people, but the likelihood of that happening is unlikely because I think people have social media just because they want to be a different person to other people, and change other people's perspective on how they're looked at." I ask Lauren if she ever wants to appear differently to other people. "No," she says emphatically. For Lauren, social media is purely a tool for staying in touch with family and friends.

But then she sighs and rolls her eyes. "I don't know why I have social media," she says. "I keep deleting it and deleting it, and then I keep reactivating it."

It turns out that this is something that many students do, but Lauren expresses extreme frustration with herself about constantly going back and forth. She seems to suffer from a kind of social media whiplash. "Some days I wake up and I'm just annoyed with all of it, so then I'll deactivate all my accounts, and then in a few weeks, people will be like, 'Oh, did you see that Tweet?' or 'Did you see that Facebook thing?' and so I reactivate it, because I want to stay involved and see what's going on on social media." She sighs once more and gives me an exasperated look. "Then I deactivate it again," she adds.

For example, she tells me, just this morning, before our interview, she deactivated her Twitter account. "I just thought, 'Man, I really hate Twitter today,' and I deleted it, but I'm sure in a few days, I'll reactivate it again." Her reason for quitting is that she thinks people express too much negativity on Twitter, and she just doesn't need any more negativity in her life.

Lauren has done the same deactivate-reactivate dance with Facebook. Her account is currently active but only because she unfriended a lot of people so she can see only the important ones. The people she unfriended don't seem to have noticed, either, which cracks Lauren up. "No one's said anything to me," she says. "So that shows how much of an effect I have on people."

Her reason for deactivating Facebook—besides the fact that she's just too busy with school and work—is that she finds it depressing. Her departures from it have lasted as long as six months. The only account Lauren has never deleted is Instagram. She likes looking at the pictures, she says with a shrug. They don't seem to bother her.

The first time she quit a social media account was a few years ago, when some girls started calling her a "slut" on Twitter. Lauren says she never cared that girls were engaging in behavior that might amount to cyberbullying—she feels pretty immune to the name-calling, and thought they were just being dumb—but it was really annoying that they were doing it in public. So she quit.

Social media, smartphones, they're a burden, according to Lauren. They ruin friendships and cause drama. If someone posts a picture of Lauren with a couple of her friends and another friend sees it and feels left out, suddenly everyone is fighting. It's just so tiring. I ask Lauren if she ever feels jealous of what she sees people post on social media, or if she ever finds herself comparing herself to others, too. "Oh, yeah, all the time," she says. "But then I realize that I'm beautiful and I am who I am, and there's nothing that can change that."

I ask Lauren what it's like after she deletes certain accounts.

"It's great," she says enthusiastically. "I don't have to be on my phone that much anymore. It's awesome. I can actually interact with people face to face. It's enjoyable to know that people can still interact without having to use social media." Lauren may go back and forth, deleting and reactivating her accounts, but she's already made up her mind about the role of social media—or the lack of one—in her life going forward. "I'm so over it," she says. "I'm just waiting for the day that I finally delete all of them, and then I don't have to worry about it, and just get a flip phone. I'm just waiting for the day where I finally get fed up with it and I just go change my iPhone in for a flip phone and just call people on the phone to get ahold of them." Lauren credits her childhood for her unusual attitude. Her parents were always pushing her and her siblings outside to play. They didn't have regular access to computers and smartphones.

Lauren has a plan for quitting all of it, the social media accounts and the smartphone too—and permanently. "I've really thought about this a lot," she says, laughing. She'll do it by the time she's a junior, before she has to worry about getting hired for jobs. Lauren has imbibed that lesson about the importance of maintaining a professional presence on social media, but rather than fix her profiles, her answer is to quit. The thought that future employers are watching her online makes her so anxious that she'd rather have no social media presence at all. Employers will have nothing to find and therefore nothing to hold against her.

"I want to be able to get a job, and if they're going to be looking at my social media sites, I don't want them to see what I follow and possibly think that that would impact who I am as a person, so they can prejudge me," Lauren says. "It has to *all* go away so I can get a job. I don't want *anything* to affect me getting a job. Because that's the goal, after four years, you have to get a job. And if social media is going to be a factor in that, then I'll just delete it."

I ask Lauren if—after she *does* get a job—she'll restart her accounts, especially because so many students seem to think that "maintaining a social media presence" is also an important professional responsibility. "No," she says, shaking her head. Then she backtracks a bit. "If I really needed another account, I'd probably just make a new one with the job, and just make it all about the job."

Lauren also mentioned her dislike of smartphones several times. She is pretty old-school for her age—she doesn't like to text, preferring to actually talk to people on the phone. Despite her dislike of the device, though, "It's attached to me at the hip," she says. She tries "really very hard" to stay off it. It's distracting to both her schoolwork and her relationships. So, I wonder, why does she take her phone everywhere? "It's an expensive piece of an accessory," she says, matter-of-factly. "It was two hundred dollars, and I really don't want to lose it or misplace it." Plus, it keeps her informed and is good in case of emergencies. Every chance Lauren gets, though, she'll leave her smartphone at her parents' house—for weeks at a time.

Leaving her phone behind provokes the same sort of joy she experiences when she quits social media. "It's amazing," she tells me. "It reminds me of what I was like before I got my phone. It was *great* before cell phones," she says wistfully, then laughs. "Yeah. It's great! I don't have to worry about whether I lose my phone or not, and I don't have to worry about anybody trying to contact me. It's just really freeing. If people need to contact me, they'll find a way, but, I mean, other than that, sometimes you just need a break from your phone, and need a break from everybody constantly trying to get ahold of you."

I ask Lauren what quitting might do to her relationships with others. It shouldn't affect them at all, she tells me. "If it does, then you have the wrong friends because friendship is based on face-to-face interaction, and you have to spend time with people," Lauren says. "You can't just sit there and have a relationship over the phone. That's not a relationship. You need to actually see them face to face and go do stuff with them, go to the movies or go and do something with your life, don't just sit there on your phone the whole time." Lauren pauses for a moment, then offers some of her last words on the subject. "I look at the bright side of things," she says. "But I feel like social media just looks at the glass half empty rather than half full."

ELISE: THE GREAT SOCIAL MEDIA PURGE

Elise, a tall, beautiful biology major at a Catholic university, finds the intense effort to get "likes" on Facebook exhausting—and disillusioning, so much so that she describes it as "a really, really dangerous, vicious cycle to get into." To avoid succumbing to this vicious cycle, and to stop having to face this incredibly upsetting behavior in all of her friends, this past summer Elise shut down all her social media accounts for a full month.

Her brother calls it "The Purge."

"I just deleted it for a month, and it was awesome," she says, with excitement in her voice. "It was *really* great. I only put it back up, coming back to school." Elise wanted an easy way to stay in touch with family from home. Currently, she's on Facebook and Instagram. But "The Purge" has had a lasting effect on Elise. Not only does she use her accounts less frequently than she used to, but she's also decided to make "The Purge" an annual, or maybe even biannual, ritual. "I'm just going to shut all my social media down for a month, and that'll be great."

I ask her why she likes this purge so much.

"It was just awesome because there were no distractions," she says. "I worked at a dentist's office this summer, and it would be so easy sitting back in the break room by myself, to be on Facebook or Instagram while I was eating." Then Elise would get home from work and do the same thing. This constant checking, looking at the same things again and again, really bothered her. Quitting, Elise explains, allowed her to focus more and to do other things she loves, such as read. Instead of being on Facebook in the break room, Elise would sit and read a book. She quickly realized how much she preferred reading to constant scrolling. Elise stopped thinking so much about what everyone else was doing while she was at work, whether they were on vacation, or what they were posting about their summers. Being off social media made Elise pay attention to her priorities. "While I was doing [The Purge]," she says, "this definitely let me focus more on what's really important to me. And, that is *not* my friend's post about her vacation. That is *not* important. Am I happy that she's having such a wonderful time? Absolutely. But she can tell me about that when she gets back."

Elise also loved not checking "fifteen times a day" to see how many "likes" she got on a photo. It *really* bothers Elise how much she used to care about getting "likes," and this was one way of breaking the habit. Elise also learned to call her friends up to talk to them instead of going on social media to see what they were up to. Her entire "mindset" about herself and her friendships changed. Elise returns again and again to the subject of how upsetting it is to see what her friends are doing on social media, and also how much it upsets her when she sees her friends in person and they're on their phones, obsessing about their posts. Elise has learned to pay attention to how her friends "operate in person," and she really hates it if a friend spends time on social media when they're together. Such people go down a notch in Elise's estimation. Now, she feels like most of the central people in her life aren't the type to be on their phones all the time, especially not when they're with her.

Elise likens "The Purge" to a juice cleanse. It has a religious analogue, too: Lent and fasting. Engaging in a social media fast, of sorts, truly felt cleansing to Elise. In fact, as we continue talking about it, Elise's sense of how long and how often she might purge expands. She begins to imagine getting to the end of year and being able to say she was off of social media for *three* months, not just one or two. Or even *five* months. Then she imagines what it would be like to get off of social media permanently. "It'll probably get to the point after undergrad, I'll probably delete my Instagram," she says. "I think after [college], you're working in grad school or you're at a job and, you know, you really shouldn't be spending time on Instagram." Social media is for "young adults," Elise thinks, definitely not adults. She might end up keeping Facebook, though, to keep in touch with her aunts in Florida. But at the moment, she's not sure. You never know, she might quit it all for good after she graduates.

QUITTING (TEMPORARILY) IS TOTALLY TRENDY

I was surprised to learn how often frustration with social media led people to quit, at least temporarily. Even those students who didn't find social media all that troubling usually considered quitting at least one of their accounts at some point or another. The urge—the compulsion—to constantly scroll through the Facebook timeline and Twitter feed bothers them. They regard it as an addiction of sorts and wish they could learn to control it. Most can't seem to set those limits, however.

Which is why quitting altogether becomes appealing.

From the student who quit Tumblr for a while because she was "afraid that it was becoming an emotional crutch," to the one who got off social media on the advice of a therapist, it's very common for college

students to take breaks from certain social media platforms, at least temporarily.[1] In fact, in the online survey, 68 percent of the students who
chose to answer a question about whether they'd ever quit any of their
social media accounts, either temporarily or permanently, said they had.[2]

Most interesting of all are the reasons students do this, as well as the
caveats they give about leaving or staying. Among the 32 percent who said
"No, I've never quit," their answers were often followed by hedging. These
students qualified their "no" statements with "buts" that explained how,
though they've never quit, they curb their usage in other ways, they've
"unfollowed people," and that's been "liberating." They've deleted "content," they use their accounts less than they used to or don't use them
much at all, and they either *think* about quitting or *plan* to quit in the
near future. So, even the students who've never quit have thought about
quitting or limiting their usage.

Among the students who have quit—some permanently, some
temporarily—their reasons are all over the map. Some feel they are "too
obsessed" or "too addicted," and they don't like "how compulsively" they
check it, wanting to "break the habit." Others quit because people are
"mean," there's too much drama, it makes them depressed, and it "was
taking over" their lives.

Students' reasons for, and their feelings about, quitting and coming
back or the possibility of staying off a particular platform permanently
are myriad. But it is clear that students often think about quitting or often
feel the need to quit (if not permanently), and many of them *do* actually
quit at some point.

Elise's "purge" was a calculated effort to enact a temporary, structured break, whereas with Mae, the young woman who was bullied, shut-
ting down her accounts was a way to escape emotional pain and anguish.
The more conversations I had, the more I heard about the many reasons
and creative methods students have for limiting their social media activity. There was the young woman who created a "Don't Touch" folder on
her smartphone where she would put her social media apps at various
times of the year—especially during busy academic times—as a way of
convincing herself to lay off things for a while, but also to avoid having to
recreate her accounts when she went back. There was the hockey player
who hadn't yet quit but who had dreams of quitting. He spoke wistfully
about how people used to memorize each other's phone numbers and how

he used to head out to shoot pucks during his free time, but now he finds himself spending time online instead of doing the things he really loves. He fantasized about the day when he'd be at a place, both careerwise and socially, where quitting all his accounts would become feasible.

Take Javier, a senior philosophy major at a Midwestern Catholic college, who had been completely off social media accounts for eight months when we met. He and his girlfriend decided to quit together. He was thrilled about it. His relationship to social media had a definitive emotional arc: it started with amusement, then led to disenchantment and eventually full-on depression, which is why he stopped. Mark Zuckerberg makes it really hard to quit, he tells me, because Facebook keeps trying to lure you back. Javier has resisted so far and says he will continue to do so. He felt like he had a "chemical addiction" to social media, and quitting was really difficult at first—it created a "hole" in his life that he needed to find ways to fill—but it helped that he and his girlfriend were in it together. To get over the addiction, Javier had to find new ways of getting that "chemical happiness," most of which involve "real-world interaction," he says, which social media had really limited. Now he uses his time more constructively, which he really likes. He studies more, gets better grades, and spends a lot of time reading the "real news."

Javier refers back to what he calls his "chemical addiction to Facebook" multiple times. The addiction, he says, had to do with posting things solely to get "likes." They meant so much to him, even though he knew they were meaningless bits of affirmation. "That's what made me depressed," he explains. "I was spending so much time working for these things that were meaningless . . . .you don't really know what you're working for. You're just working for the next flag to pop up, but after that, then it's like the same exact goal all over again, and nobody really cares beyond you. So I'd say it was a relief [to quit]. Maybe chemically there were some things lacking when I finally gave it up, but intellectually I was like, 'Oh my gosh, yeah! I hate Facebook!'" he adds, laughing. Javier has all kinds of strong feelings about social media. He says it made him judgmental of both himself and others, and he didn't like this at all. He felt bad about himself and others nearly constantly.

There's a downside to not having it anymore, of course. Javier is a little clueless about events on campus and what his friends are up to. Sometimes he doesn't get invited to things because people do all their inviting on Facebook. Despite this, he's upbeat about his decision. "In my heart, I'm happier getting left out than getting brought into that mess, just because of the way it made me feel," he says. "I don't think it's worth it at all to get invited to a party just to feel that horrible about yourself." Some people on campus find it confusing to meet someone who doesn't have Facebook—now that he's not on social media anymore, he can see how addicted other people are. "I would strongly advise people not to spend a significant portion of their life on Facebook or social media," he says.

When I ask Javier if he thinks he'll ever go back, he shakes his head. "Yeah, I don't ever want to go back to it," he says. "I think I'll just send emails to people and meet with them in person."

Amy, a senior at her public university, explains, *emphatically*, that she considers quitting "all the time." When I ask her to clarify (does she quit and get back on, or just think about quitting?), I find out that, like Lauren, Amy *was* "quitting all the time," though now she's "found a way around [this]."

"Every so often, I would deactivate my Facebook because I got sick of seeing what everyone was doing," she says. "But [eventually] either I would want to Facebook stalk someone, or post pictures, or see pictures, or see someone, something like that, so I'd always go back. So I would disconnect for a little bit, but it would always suck me back in." The work-around Amy figured out allowed her to "quit" only the parts that both-ered her. "I wound up unfollowing everybody on Facebook, so when I log into Facebook, *nothing* comes up on my newsfeed. So I can post my pic-tures, and I can see when I'm tagged in pictures, and I can go and look at my friends' walls and write on their walls, but it's not thrown in my face, everyone's events and pictures and thoughts and things like that. So, it's working for me now."

I ask Amy why she likes this setup better.

"Everyone's all up in your face about their opinion and what they're doing," she responds. "I think it's easier if I can pick and choose what I see, and especially if I'm just looking at my close friends, because I'm interested in what they're doing and what they're thinking. It just kind of waters down the whole thing for me."

The issue of "picking and choosing what we see" is a theme that comes up often when students tell me why they try so hard to limit access to their accounts, or even quit for a time. Feeling left out, experiencing hurt when looking at someone else's timeline, and getting stressed or sad that everyone else's lives are better than yours are among the most common reasons why people quit or curb their usage.

This is certainly the case with Hae.

HAE: IS FACEBOOK USING *YOU?*

Hae is a shy, first-year student at a Christian university in the West, who shows up for our interview wearing a suit and heels. She's serious, studious, "100 percent Korean," she tells me, and very devout in her Christian faith. One of the most difficult things about college for Hae is learning to balance her time. She has become overwhelmed trying to juggle her studies with her other responsibilities, plus trying to find her way socially. Lately, Hae's been struggling with the meaning of life and what (if anything) that has to do with academics and grades, something at which she's always excelled. She doesn't want grades to rule her life anymore, she says with a heavy sigh.

Hae is on both Facebook and Instagram—but only because college practically requires them for communication, in her opinion. She's just coming off of a four-month social media hiatus. It turns out that Facebook and Instagram were getting in the way of life's meaning. In high school Hae started going on mission trips and realized that it was the face-to-face interactions she had with people—and *not* whether people are "liking" her photos on Instagram—that made life feel worthwhile. Hae got tired of "hanging out with people always trying to take pictures of the moments of fun that [they're] having, and trying to post them, and seeing if [they] can get one-hundred-plus likes and seeing who comments on it, who doesn't."

Hae tells me that she deleted her social media accounts during her senior year of high school. "I was a lot happier this summer without [social media]," she says. "I didn't have to think about my old friends who were going to different colleges, and I had a select group of friends that I would talk to, and so with everybody else, I didn't need to see their partying pictures over the summer and didn't need to see the concerts that they went to. I could just kind of focus on myself and spend time with my family and just enjoy my summer without those kinds of social pressures in the back of your mind, like oh, I'm here and it's a Friday night, at home with my family, and there's other people who are obviously out doing other fun things that I'm not." As with Amy, not having to feel left out is a big reason that getting off social media was a relief, but Hae also says that it helped her be "more present-minded" and it allowed her to focus on herself for once.

Hae's friends were shocked when she quit. But when they protested, she told them she had to "because I hated it." Her friends were bummed they could no longer tag Hae in things, but Hae liked that they couldn't. "[Social media] just consumed so much time and energy and emotions and I just didn't want it to govern my day," Hae says. "Or the way that I feel on a particular occasion because I wasn't invited to something and I saw pictures of all my friends."

I ask Hae if there was a specific event that provoked her to delete her accounts. One reason, she tells me, was that she'd gotten her Facebook account in middle school, and she had too many "friends" who weren't friends at all. Hae didn't like that she was spending so much time paying attention to the lives of people who weren't part of her life in any real, meaningful way. "The other is because I just knew subconsciously in the back of my mind that as a girl it's really easy to be insecure," she explains. "And I have been insecure in my life. I just didn't feel like I needed it. I just wanted to delete it for that reason. Just to liberate myself."

Hae is "kind of sad" that she has reactivated accounts, but even sadder that she "only lasted four months" away from social media. What makes her saddest of all is that not being on social media is nearly impossible. She only went back because so many people at her college (including professors) use social media to communicate. Hae was also having trouble meeting people when she first got to campus. "I feel like it's sad that social media has become such a big part of people being able to develop relationships, but it is something that I think has become so ingrained in our society, it's just not going to go away any time soon."

Now that she is back on social media, Hae is doing some things differently. She started over completely—she didn't just reactivate her pages. She doesn't accept friend requests from people she doesn't really know, and she accepts requests only if they are people she wants to see on her newsfeed. Hae has made it so that she's looking at posts only when she wants to see people they are doing. This is also a way to protect her privacy. She doesn't want anyone who isn't close to her reading what she writes or seeing her photographs.

Hae is also more conscious about how social media can really do a number on a girl—and she's vowed to not let this happen again. "I think it does mess with you mentally," she says. "I realize now, looking back, how the Internet is not a safe place to just openly say what you're feeling and to openly post pictures. . . . I just don't like the idea of not being in control of who knows what I'm thinking and who knows what I'm saying." Hae is doing everything she can to maintain control of her accounts now, to not allow them to make her feel hurt or out of control. She's trying to use Facebook and Instagram only in ways that are useful to her, and that serve her current needs either socially or academically. In the best way she knows how, Hae is struggling to use social media without losing what considers meaningful to her.

KATIE: IMAGINE A WORLD WITHOUT SOCIAL MEDIA!

Katie, a senior at her public university, is sweet and reserved as she patiently answers my initial questions. When we start talking about social media, however, Katie's entire demeanor changes. She becomes passionate. Katie deplores posting all the time—she feels like this is a trap people fall into. And like Hae, Katie is worried that social media is sapping the meaning from our lives, or at least from hers. "I don't want to end up feeling like things I do don't have meaning unless everyone else knows about it," she says.

Katie didn't always feel this way. She used to love going on to Instagram. She enjoyed the constant stream of photos from people she knew. But she quickly became disillusioned with it—and with all social media. As Katie scrolled through everyone's updates, she began to realize that people post only "the bright side of things, but never bad things." This lack of honesty, the distortion of reality, really bothers her. "I just feel like it makes people seem fake," she says. Plus, it stymies communication with her friends. If everybody sees what's happening with everyone else on social media, then they feel like they don't need to make personal contact anymore. On one level, this fits conveniently into everyone's very busy lives. But on another, Katie has realized that just because people appear happy on social media doesn't mean that they are happy in real life—and friends going through tough times can end up feeling isolated and alone as a result.

Katie has quit her various social media accounts in the past. She got rid of Snapchat at one point, though then she felt left out because her friends were constantly "snapping" each other, and she wouldn't know what was they were talking about half the time. So she went back. "Then I've gotten rid of all of [my social media accounts] at one point last year," she says. "When finals came, I was like, 'I don't want any distractions. I want to just be studying, not worrying about stupid stuff.' So I just got rid of it for, like, three weeks, almost a month. At first, I'd take my phone out and go to check on social media, and be like, 'Crap, I deleted the apps.' But after a while, like, after two weeks, it kind of became regular." Once Katie got used to not checking her social media apps, she enjoyed not having them. She began interacting with people in person more often. She focused on what was happening with herself more than what everyone else was doing.

But when winter break started, Katie downloaded everything again.

It's one thing to quit for a while; it's another to stay off permanently. Katie believes it's very hard to maintain an active social life if you are off of social media entirely. She resents this, however, and wishes she could quit altogether, forever. Katie thinks about quitting all the time, and her boyfriend fully supports this. They go back and forth a lot, discussing whether they should take the plunge.

But they never do it.

Katie *really* does not want to feel like she's missing out on what everybody's doing, though she suspects—she *hopes*—that this will change as she gets older. "I feel like the older I get, the less and less I care about it. So maybe eventually, I'll just get rid of it, but not really worry about it."

Katie longs for a world entirely without social media. She thinks life would simply be better if it did not exist. As she elaborates, "I think that if everyone didn't have it, life would be a lot better, but the fact that if *I* just got rid of it and *everyone else* had it, it wouldn't be that much better because then I would just feel left out."

Katie hopes that social media will have disappeared by the time she has children. "I hope by then it's gone," she says with a laugh. "There's no need to say everything that's on your mind. You don't need to post it to feel like you've gotten it off your chest. You don't need to have self-validation from social media. You don't need to post things to feel good about yourself or get a certain amount of 'likes' to feel good about yourself," she adds. Katie's wish for her future children sounds like a wish for herself.

ARE THE GOOD THINGS GOOD ENOUGH?

There is an underlying angst among college students about how much time they spend on social media, as well as how some of these platforms affect their feelings about themselves and others. Many students engage in a cycle of quitting and returning, then regretting it.[3] For some, quitting simply reveals their ambivalence about the medium, but for others it belies a high level of stress and a desire to escape the pressure and unpleasant emotions that come with social media.

Regardless of which category they fall into, the young women and men I interviewed and surveyed are searching for *rules*.

Right now, social media is a vast and largely unregulated sea of information, emotion, drama, competition, showing off, pressure, expectation, and a million other things. It's still incredibly new, in its infancy, really, and society has yet to catch up to this dramatic change in our world and how we live in it (or escape from it). We've yet to truly begin to unpack what it means for humanity, ethics, the construction of self, gender, race, our socioeconomic situations, and the enterprise of education. And this leaves young people floundering.

At the moment, many of us are social media gluttons, unable and often unwilling to curb our appetites, trying to navigate through a never-ending battle of wills against our brain's desire for that quick fix amid a consumer culture determined to stoke our addictions with ever more clever platforms and devices.

Amid this, young adults are searching for a footing somewhere, a way to anchor themselves in this sea, because too much of anything, including something we love, can eventually make us sick. Students are searching for rules and, not finding them, are making up their own.

They want boundaries. They want guidance. They want to talk about their experiences of social media and their struggles with it, to try to figure out how to better live with this new force that has taken over our lives.

They want to be "liberated" from it, in the words of several, in the sense that they feel *controlled* by it and even *used* by it, and they want to regain control.

The genius of social media is that the more people join a network, the more difficult it is for others not to join it. This has made Mark Zuckerberg and others a lot of money, but it has also made a lot of people miserable. Social media has become so embedded in our lives that quitting can make social—even professional—life impossible, especially for younger people. Even if they work up the nerve to quit, and are much happier as a result, and even if they aren't sucked back in by simple curiosity, it is very difficult to live your life—especially college life—without it. When everyone communicates by Facebook, when everyone is expected to be on social media, it becomes a burden not to be. And usually that burden is too much to bear, so even the students who quit for a while, eventually go back.

Plus, there is plenty that makes the allure of social media difficult to resist.

By far across the interviews and in the online survey, when it comes to what young adults love most, connectivity is everyone's number one, favorite thing. Of the students who answered an essay question about the best thing social media offers, 43 percent identified the way it allows us to connect with, keep up with, and stay in contact with old friends and family members that are far away, in addition to being a wonderful tool for making plans.[4] A distant second were the 12 percent of respondents who cited sharing and seeing other people's photos as the best thing about social media.

These good things would be difficult for most people to give up over the long run.

A few students couldn't find anything nice to say, though. They began their answers with "I hate it," and went on to discuss why. Also notable was how for many students, listing their favorite thing also provoked them to reveal the things they hated most. The connectivity and sharing can be a double-edged sword. Several students mentioned in the same sentence that what they loved the most was also what they disliked the most, as with the person who wrote simply and briefly, "Favorite is connectivity, least favorite is connectivity."

"My favorite thing about social media is you can be as anonymous or as public as you want," wrote another student. "And my least favorite thing about social media is you can be as anonymous or as public as you want." One young woman began by talking about how great social media was for photos, for finding out birthdays, and quick access to an overview of someone's basic interests. But after listing these pluses, she finished up harshly. "It takes away the personal conversation. While it can be convenient I think social media is destroying our generation. We cannot hold a conversation, we don't have an attention span longer than six seconds, and we are obsessed with ourselves." There was also the young woman who spoke of the wonderful photo-sharing aspect of social media, and then went on to complain about just this. "I don't like how addictive it has become and how I now live my life in terms of taking pictures for social media," she wrote. "It's taken away just being and living."

Several students used this question as another opportunity to complain about the appearance of happiness. "I don't like how the purpose [of social media] for some people is only to feel validated by others. . . . [P]eople usually only put their best foot forward which creates an illusion that everyone is happy all of the time," wrote one of these respondents. "Everyone always seems happy and people seem to have the perfect life on social media because they only post about good things that happen," wrote another. "This can make others feel depressed or that something is wrong with their own lives because everyone else seems to be having such a good time."

A separate essay question asked if students had any advice they wish they'd received when they were younger or that they would give to the next generation. These responses were full of guidance about getting accounts later in life, counsel about not posting so much, and warnings that what goes up on the Internet lasts forever. Many students commented that they wished that someone had told them not to care at all about social media, that it doesn't really matter, and that the important things in life have nothing to do with Facebook, Twitter, and Instagram. And there was a lot of advice about not comparing yourself to others, being aware that social media is fake, that people lie, and that people are not showing you their real selves in what they post.

The students who seem to handle social media best are the ones who are able to be ambivalent about it—those young women and men who can manage the self-promotive dimensions without too much stress, who can live with the pressures of constant evaluation, and who aren't made so emotionally vulnerable by social media that its negatives wreak havoc on their self-esteem.

Apathy has become a healthy mode of survival.

But is this the lesson we want people to learn: that the best way to deal with one of the most significant aspects of our social lives is to master an ambivalent attitude about it? Is learned ambivalence the way we want people to cope with the changes that social media and smartphones have brought us? Is this, in short, the best we can do?[5]

The alternative is for everyone to remain vulnerable to social media's ups and downs. Even if young adults work hard to mask their feelings, the 24/7 stream of social media and the ever-presence of smartphones become a roller coaster of banality, plan-making, fun, disappointment, stress, hope, pride, loneliness, distraction, showing off, pressure, and a million other things. On the surface, this sounds very human. Any social situation can be wonderful and fun or stressful and awful, and everything in between. Life is messy and angst-ridden, full of unexpected potholes, words better left unspoken, and painful disappointments, just as it is full of joy and love and those moments that no one ever wants to forget.

Social media reflects this reality. But it doesn't *just* reflect this reality. It adds another dimension to it, enhancing this reality by making it public and constant.

Because social media is so public, because every post has an audience—and potentially a very large audience (not to mention a very large *judgmental* audience)—it is like a hyperintense form of reality, a pressure-cooker version of life itself. It's no wonder that so many students take regular breaks from it. And while it's true that people can use social media to broadcast their best achievements, it's equally true that their most embarrassing moments might be visible on an enormous scale and may, in turn, follow them forever. The sense of permanence and high-stakes consequences for one's actions can overwhelm some people and paralyze others. And everyone is learning to watch their steps, often to an extreme. What began as something intended to be social, expressive, and fun has become a burden to many, none more so than the young who now face growing up online. And the professionalization of social media and the expectation to appear happy are exacerbating a success-driven and overachieving culture during high school and on college campuses.

Despite these drawbacks, very few people are interested in giving up the benefits of social media. It has come to dominate our lives *because* of its benefits—the way platforms like Facebook make it easy for us to stay connected, make plans, and record memorable moments. These advantages are too appealing for us to go back to the way things were. It's unrealistic, too. Social media isn't going anywhere, so the question becomes:

How can we better live *with* it?

And how can we better advise the younger generation on their use of it?

CONCLUSION

Virtues for a Generation of Social Media Pioneers

Our culture today is forming around "being happy," and although that is good, I feel people actively neglect the fact that life has ups and downs. Therefore, social media is used only to highlight the ups of life, while the downs are more often internalized behind the walls of our bedrooms, homes, and personal lives. Although I do not feel that social media is a place to air negativity, I think it is okay to not be 100 percent happy all the time, and social media promotes the latter to the extreme.

> Jake, sophomore, evangelical Christian university

People want to see others as happy, and people are easily bothered by someone who confesses that they aren't happy or aren't what everyone wants them to be. If more people stepped out of their boxes, found their true selves, and posted that self online, they would get a lot more hate and they would be a lot more vulnerable, but ultimately maybe more people

would start being honest.

        Alice, first-year, evangelical Christian university

PIONEERS OF A VIRTUAL WORLD

The Facebooks and Instagrams of today are like the grand new boulevards of still nascent modern cities: public spaces for everyone to stroll
in their fashionable best, to parade their romances and their families and
their finest, to display with pride their riches and achievements, to inspect
and acknowledge the similarly garbed and the lucky, as well as look down
their noses at the less fortunate and even turn their backs on those they
simply do not want to see. Yik Yak and other anonymous platforms are
likewise the proverbial mean streets, the dark alleys and dangerous cor-
ners where all the ugliness ensues, the kind of behavior best shielded from
the prying eyes and bright lights that shine down over the boulevards.[1]
The world of social media, the new, public, and wide-reaching virtual
space it has created for all of us, is at once as old as the town square and
its corresponding gossips, and as new and vast, as unpredictable and dangerous, and as thrilling and full of pontential as an uncharted continent.

    And now we have entire generations growing up in this wild terrain.

    Seen in this light, the young are not so much narcissitic. Rather, they
are the new explorers, brave and courageous, testing out the unknowns,
both the good and the challenging, succeeding and failing and doing
their best to at the very least survive, so they can pass on their hard-won
wisdom and advice to the next generation so *their* experience might not
be as rocky, and so they can better live with social media than the previous
generations. Young adults of today have little choice other than to build
their lives in a virtual sphere that did not even exist (or only barely) when
their parents were children—a fact that can boggle the mind. They—and
we—have no precedent for a life lived and celebrated and picked apart on

a virtual scale. Yet the children coming up are mapping out their lives and experiences in this unbelievably new sphere regardless of its unknowns, because it is where our world is headed, and where their lives have begun.

And soon, growing up online will be *all* anyone knows.

But *this* generation is the test generation, the one that faces working out all the kinks and complications, while we—their parents, coaches, teachers, mentors, professors, admissions officers, bosses, and future employers—are likewise faced with helping them through this massive cultural shift as best we can.

What I have called the happiness effect throughout this book—the requirement to appear happy on social media regardless of what a person actually feels—is an effect of our own making. We are the ones who have created this problem. Young adults have internalized the lesson that if you can't say something happy, you shouldn't say anything at all, even if you feel despair, dismay, anger, or any number of other emotions common to human experience, *from us*. We have burdened them by obsessing about how people in power might react when confronted with evidence that sometimes we are silly, do stupid things, get angry, say something dumb, appear less than perfect, and maybe even drink a beer before we turn twenty-one. This lesson on our part is obviously well-intended and, at least on its surface, sounds like excellent, rational advice. But the consequences are disturbing. Posting on social media for so many young adults means pretending one's true feelings are not really there; it requires hiding them and, ostensibly, lying for the sake of one's audience. Because of this, most of what anyone ever sees on social media are gleeful timelines of joy and accomplishment—the highlight reel. This can make anyone who isn't blissfully happy all the time feel even worse.

And none of us are immune to *this* part of the happiness effect—not really. No matter what age we are.

While reflecting on what I've learned from talking to college students about the state of social media in their lives, the happiness effect and its corresponding highlight reels, I've found myself thinking about the photo albums I made when I was younger. How I would sort through my pictures to find only the best ones, because of course I wasn't going to memorialize a photo in which my friends and I looked terrible, and I certainly didn't showcase the awful times in my life. There were no pictures from funerals (but then, nobody took pictures at funerals either), no captions saying that just moments before I went to a party where a photo was taken, I had been weeping about a boyfriend who'd broken up with me.

In other words, my old photo albums are highlight reels, too.

So then, how is what young adults are doing (and what we've taught them to do) on social media any different than the photo albums of my youth? Couldn't we simply regard it all as the same, except that their photo albums are digital? Which would render our advice to them rather harmless in the end?

Yes.

But also no.

First, I wasn't constantly making photo albums. It was something I did maybe twice in high school and once during college. Second, although I definitely enjoyed having photos of important moments like prom and graduation, and maybe a fun party with the favorite people in my life, I wasn't constantly taking pictures of everything and everyone, everywhere I went. It wasn't practical. There wasn't an expectation of near-constant documentation. What's more, the photo albums from my youth are full of silliness and ridiculous antics because those were the fun moments I most wanted to remember. There are plenty of pictures of people doing things that our future employers might have frowned upon. I didn't filter what I put in my albums based on others' expectations and nobody imagined I should—because I made them for *me* and for the precious few people with whom I decided to share them. My photo albums were private possessions, and I could control who saw and didn't see what was in them. The possibility that anything in them would go viral did not yet exist. I always knew who saw my albums because the viewers were with me at the time—I had to personally grant them access.

Would I have brought one with me to a job interview? Absolutely not! It would be *none of the interviewer's business*. What's more, I wasn't about to hand over my album of happy memories to a crying friend. Would *anybody* with a single compassionate bone in their body do such a thing? I hope not. That's not how you console someone you care about. I also didn't compulsively check my own photo albums and the photo albums of others every day. For someone to share a set of photos with me was a rare occurrence. I'm glad I have those albums from my youth, but I haven't looked through them in years.

When I was in college, before cellphones were widespread (and long before the existence of smartphones), cameras and video cameras were rare birds that one saw only at the most significant formal dances and in the hands of parents at graduation. This meant that what happened on campus largely stayed on campus and in the fuzzy memories of those people who were there to witness it firsthand. In a recent conversation with some college students in California, we ended up talking about the age-old tradition of streaking—running naked across the quad. It's not that streaking doesn't still happen, but during that class, we had a conversation about how, when I was their age, you could pretty much run naked all over the place if you were feeling so inspired. Photos or videos of our bodies would never go up online, since this possibility didn't yet exist, and so we had nothing to fear. And while cameras *did* exist, if someone had taken one out to snap photos, they would have been judged harshly for it. You just wouldn't do such a thing. Today you can still streak, sure, but the consequences could be dire. I use streaking as an example because it's so typical of old-school college fun, yet today's students don't have the same freedom to act on the playful impulse to run a little wild. They *can*, but doing so might cost them their future and lead to tremendous humiliation—and we make sure they know these risks to their very cores.

I have often wondered how my own experience of high school and college would have been different if I had constantly been creating photo albums with captions to show to everyone I knew—including my grandfather and future bosses. What if I'd been constantly aware that any little thing I did might end up memorialized in a photo for all time, and one that—in the worst of circumstances—could go viral and ruin my future prospects? It's impossible to know what would have changed, but having talked to so many college students about how social media is affecting the way they conduct and document their lives, I can get at least a general idea. And I've tried my best to paint that portrait here.

Because of social media, there is a major difference in the *constancy* of creating content, the *kind* of content we choose to create, and the *control* we have over that content and who sees it. Social media is a 24/7 performance that requires everyone to take acting lessons. Everyone learns how to behave for their various publics, lest they get bad reviews (or no reviews at all) and risk social and financial ruin. Those of us who've already graduated from college, who have jobs and significant others and children, are also subject to these changes, since social media is not just for kids. Nearly everyone I know is on it, often all the time. I have friends who've posted thousands of photos on Facebook, who compulsively check, who worry about "likes" and putting on a "good face" for the public, just like the students do. But though the stakes can be high for anyone, they are different for those who are still very young and just learning to find their way in this new world, who are vulnerable and emotional and trying to figure out life and who they want to be when they grow up—*as* they grow up. These things were hard enough before social media; adding a potentially limitless audience of constant evaluators makes them infinitely more difficult.

This is what leads me to pause when the accusations fly about the narcissistic tendencies of this generation. As I said before, I don't think they're narcissistic. I think they are pioneers, and being a pioneer isn't easy.

Social media has changed our lives—all of our lives—quickly and dramatically. It is an incredibly young phenomenon, yet its influence on the way we live now is stunning to behold. That it is still so new means that we haven't yet had much of a chance to reflect on how it is changing our world, for good and for ill. We are still mostly reacting to as opposed to carefully responding to and shaping its presence in our lives and the lives of those we teach and mentor and parent. Yet the college experience is supposed to foster, not suppress, free thought. The notion that the best way to handle all posts—especially if you are a college hopeful or soon-to-be college grad—is to make them positive and happy is a *reactive* teaching. It's triage, a survival mechanism designed on the fly to cope with the incredible invasiveness and pervasiveness of this medium. Social media has the potential to be incredibly destructive if the wrong post or photo gets out, the thinking goes, so the best course of action is to protect yourself. This teaching is a corrective to all those college students in the early twenty-first century who posted photos of themselves doing keg stands, smoking joints, being horribly vulgar and politically incorrect, and it keeps people from being overly vulnerable and overexposed to an enormous audience of not only friends and acquaintances but strangers, too.

We all need to take a step back and stop reacting for a moment.

As a culture, we need to take time to rethink the advice and teachings we are passing on to this pioneer generation, which will become the same advice that future generations will receive eventually as well. I do not believe we have given enough thought to the effect social media is having on our capacity to be vulnerable, to be emotional, to be human and therefore imperfect. Or to the fact that we are living in an age where young adults are growing up learning to think of themselves as brands and as marketable commodities. We need to ask ourselves: Do we want to raise employees, or do we want to raise children? Do we want colleges to create "brands" or to support emerging adults eager to become good critical thinkers and citizens of the world? Or, does the dominance of social media mean that to be a citizen of the world today *is* to be a brand? Is this simply a reality we (and our children and students) must accept or be left behind? Is the "appearance of happiness" the *result* of good critical thinking about social media? Or is the cost of *not* always appearing happy, the price of *honesty*, too great for us to even contemplate?

In short, can we become better consumers of social media so that it doesn't consume us?

EIGHT VIRTUES

Aristotle's *Nicomachean Ethics* is one of my favorite things to teach because his ideas about virtue are so practical and useful. He offers a framework for ethical decision-making that varies from person to person. For example, to drink "virtuously" will differ based on height, weight, food intake on a given occasion, and overall tolerance. For Aristotle, vice hovers at the extremes. We must work to find the "mean" in our behaviors, for it is there that virtue awaits us. And once each of us finds the mean, we must "practice" it until it becomes second nature or "habit."

Over the last several years, I've often wondered what advice Aristotle might have about social media and the ways our devices have come to dominate our lives. Drawing on the lessons I have learned from doing this research, I have devised a list of eight virtues for the social media age in the hope that they might help us navigate social media and smartphones in healthier, more critically aware and empowering ways—virtues that might help to mitigate the happiness effect, and provide a framework for rethinking the advice we currently give (and receive) about how to be online.

I. THE VIRTUE OF VULNERABILITY: HONORING OUR THINNER SKINS

While it is true that the world can be a cruel and violent place, it is not only this. To constantly brace ourselves against this darkness hardens us in ways from which it is difficult to recover. I live in New York City and am a very practical person, which has helped me to accept a certain level of cold reality. But I want to remain a person who trusts, who sees the best in others, who gives the benefit of the doubt, who takes risks and cultivates a rich emotional and personal life. Were I to try to make myself impenetrable to hurt and disappointment, a huge part of the joy and happiness I have in life would disappear. I need a thick skin to survive—we all do—but not one so thick that it is impervious to joy as well as sadness.

Having spoken to so many young people, I worry about their "skins" and what they are trying to do to them. Many of them are terrified of the exposure that can come from social media, from making themselves available for evaluation and scrutiny online. They fear that putting themselves out there might result not only in negative reactions but in *no reaction at all*, making them feel irrelevant, invisible, and insignificant. Meanwhile, the most vulnerable among us are prone to devastation and unbearable pain. Everyone is hard at work trying to steel themselves against being hurt, and when it happens anyway, we rebuke ourselves for it. On one level this is smart. It's not a bad thing to learn to protect ourselves. It's a fundamental part of growing up and learning to be in the world. It's necessary for survival.

But we can go too far and end up sealing ourselves off from beauty and joy and real connection. We can become too ambivalent. The extreme public nature of social media, along with our somewhat uncontrolled and compulsive use of it, exposes all the fault lines and fissures of our vulnerable, tender selves. Before we wall ourselves into fortresses in an effort to endure this, we must consider what we might lose. We should want our skins to be *just thick enough* to survive, and not a millimeter thicker.

2. THE VIRTUE OF AUTHENTICITY: PRIZING OUR REAL SELVES OVER THE VIRTUAL ONES

The phenomenon of comparing ourselves to others, coupled with the pressure to "keep up a happy appearance" on social media to prove to everyone that you matter, that you are having a blast, that you have friends and achieve enviable things, might seem like it would turn everyone into little narcissists. Yet the students I interviewed and surveyed weren't egomaniacs. They seem, more than anything, deflated about their real lives. Their self-esteem is far from sky-high.

College students care that they see their friends acting fake online, and that to keep up with the times they need to act fake, too—or at least filter what they post to the point that it reflects only one sliver of who they really are. The popularity of Snapchat and Yik Yak embodies students' desire for honesty and authenticity on social media, yet they are learning that true authenticity requires anonymity, or at least impermanence, because they are afraid to be honest and truthful on the platforms attached to their names. Standing up for who we are and what we believe is something that young adults are learning is best done anonymously.

What's more, the anonymity of apps like Yik Yak encourages sexism, racism, and an excessive level of meanness and cruelty, rather than empowering young adults to find ways to productively express their real feelings, selves, and opinions in public. Anonymous platforms encourage self-expression, but what young people see among their peers often horrifies them. Young adults need to think about authenticity—where and how and with whom it is possible to practice it—so they can gain a better sense of how to express themselves both online and off in ways that make them feel healthy and secure. They need to think about the costs of performing "success" according to social media's ideals, rather than their own. And we need to help them think through this.

The happiness effect undermines authenticity. Young adults are spending an inordinate amount of time producing and poring over fake, idealized, public versions of themselves and others, created and stylized to stand up to the scrutiny of many different audiences. Rather than directing this energy toward self-discovery and discovery of the world, they are spending it conforming to others' desires. We all become excellent people-pleasers, without concern or worry about the effect such constant people-pleasing might have on self-development and happiness. Social media is creating a generation of young people who've learned that self-emptying and self-sacrifice for the good of *getting*, of *pleasing*, of *pandering* are more important than attending to and learning about one's own needs and the real needs of those around you. Students care deeply about authenticity, yet they also know that social media is the last place to go to find it.

3. THE VIRTUE OF OPINION: TOLERANCE OF DIFFERENCE AND DISSENT

After listening to so many students discuss the importance of Facebook Cleanups and not being the least bit provocative for fear of saying something with which another person (especially a person in power) might disagree, it's difficult not to wonder whether this phenomenon is related to the uproar on campuses across the nation around diversity and intolerance, cultural sensitivity and insensitivity, and safe versus unsafe spaces. Controversies over culturally insensitive Halloween costumes (for example) have become rampant, and there is incredible tension regarding expressions of dissent from majority opinion, or one that is deemed "politically correct." America's campuses are meant to be spaces where intellectual dialogue and debate can flourish, places where all students can thrive and live free from the fear of discrimination. In order for students to feel safe, it is important to call out racist, sexist, and culturally insensitive activities and comments—not just in person but online, especially given what people are seeing on Yik Yak.

"Neutrality" was a frequent theme among the students I interviewed. Yet there is a difference between racist, sexist, and intolerant speech, and holding an unpopular opinion. I am certainly not advocating for "unsafe" spaces, but I see parallels between the policing of speech and the overwhelmingly common belief among college students that there is an acute "danger" or "risk," with real consequences, to sharing an unpopular opinion on social media, so it is best not to express any opinions at all. I want to be clear: there is a difference between political speech and racism. But what we consider acceptable political speech changes so quickly that today's innocent remark becomes tomorrow's hate speech. Students fear that even the slightest error can—and surely will—haunt them forever (and in today's world, it might). This fear is buffered by stories of public shaming and of ruinous posts that have ended careers, destroyed social lives, and led to death threats. Students are acutely aware of how easy it is to offend people, so the best course of action is to post only "safe" material like puppy photos and smiling, happy pictures.

Are young adults learning to become so careful in their online activities that the Facebook Cleanup is leaping from the virtual sphere into the very real space of the college campus? Are students getting so good at "cleaning up" and staying neutral that any sign of provocation from a real person in real life must be squelched, even deleted, as well?

To conform to the expectations of future employers, mentors, and other people in power seems to require the appearance of perfection, and a complete lack of opinion, too—or so young people are learning. Holding an unpopular political opinion might cost you a job. Making a poorly worded statement might do the same. Rather than talk about this, or try to have a conversation about why it might be problematic or hurtful or even offensive to a particular population, we instead mount protests and call for resignations. Dialogue with anyone whose opinions do not fall into the majority has become nearly impossible. Or at least very, very risky.

Because of social media, college students are learning to "practice" fitting in to an extreme degree. And how *does* one tell the difference between an unpopular opinion and racist, sexist, culturally insensitive speech? It is difficult to parse through answers to this, when *all* speech has become such a minefield on campus. True tolerance of diversity should not result in simply conformity to the majority's attitude, and tolerance is a value we are supposed to be teaching on college campuses. But even beyond tolerance, I think we need to consider the possibility that *opinion itself* has become problematic, at least if it does not conform to perceived campus norms. We must reckon with how the professionalization of Facebook and the avoidance of posting religious and political speech is contributing to the rising fear of expression on college campuses overall.

4. THE VIRTUE OF FORGETTING: LETTING
MEMORIES FADE

Not every moment needs to be recorded and documented. There are, of course, memories that everyone wants to savor, precious people in our lives whose images we want to save, friendships and loves that photos will help preserve. But the increase in taking photos and videos simply because the tools are available, portable, and ubiquitous seems to have reached a problematic tipping point. When the point of a party or an outing with friends becomes the photo itself and not the outing and time spent with friends, we should rethink our actions and intentions. When the point of our lives becomes *proving* what we do, we need to re-evaluate.

One of the difficulties of life is the elusiveness of the moment, of our experiences, of memory. People come and go in our lives, and this can be sad. Many of us are nostalgic and sentimental creatures, and photos and videos are wonderful tools for hanging on to the things we don't want to forget and the people who mean the most to us. They are a gift in this way.

But there is a freedom, too, in the passing of time and the fading away of many experiences. It liberates us to move forward, to let go, to create new experiences in the present. There is a freedom in knowing that certain experiences will not follow us forever. Just because we *can* capture everything on video or in a photo does not mean that we should. We need to become better at discerning when and to what extent we use today's tools for capturing not only *the* moment but *all* moments. Not documenting everything allows us to live differently and do things that we otherwise wouldn't, since not everything we do should follow us forever. We need to learn to pick and choose between what we keep and what is better left behind.

5. THE VIRTUE OF THE NOW: LIVING IN THE MOMENT

Likewise, holding off on at least some of those selfies helps us to be present, to live in the now, and to experience the world more fully. College students are at once annoyed and exhausted by all the documentation they see among friends, peers, and family, as well as by the expectation to participate in that documentation themselves. There comes a point at which it is no longer fun to memorialize our activities and our time with friends, when the documentation begins to detract from and even destroy our time together. Just about everyone can speak of a time when they've been to an art museum and seen other visitors who, rather than look at the paintings and sculptures, snapped photo after photo of all the artwork, seeing it indirectly through the camera.

Collecting "proof" of every experience, every person we see, everything we do, not only takes away from those experiences, people, and activities but sets us up for another vicious cycle of doing things not necessarily because we want to or because they are fun or because they make us happy or because they mean something to us, but simply to be able to show that we did. The *quality* of experience is beginning to take second place to the *quantity* of proof that we collect. Students I interviewed and surveyed expressed tremendous resentment about this.

Learning *not* to take out the camera not only is a worthwhile skill but also will help everyone to attend to the world, the people, and the beauty around them, as opposed to worrying about the perfect shot, and then the next one, and the next.

6. THE VIRTUE OF PLAY: WHY GOOFING OFF IS IMPORTANT

Many people of my generation and older who are now parents bemoan the loss of unsupervised neighborhood play as we knew it—building forts in the woods, playing tag, roller-skating in the middle of the street, and running around all day until the sun went down and your mom or dad came out on the front steps to yell your name in the hopes that you might hear and wander home before dark. Today, playtime is highly scheduled and largely determined by parentally arranged play dates, and children are more accustomed to structure and scripted play that often involves a heavy dose of electronic devices.

Regardless of where and how play happens, it's an important part of all our lives. Just because we get older doesn't mean that our need for fun, laughter, and relaxation goes away. Even though teens and young adults are meant to act more "grown-up" and mature, playing around, goofing off, being ridiculous, wandering a bit, and being aimless sometimes is necessary for one's overall well-being. How else can we handle the added responsibilities that life sends our way as we get older?

Social media is yet another sphere of life that students and young adults are ceding to more serious pursuits like college and a career. The era of university-bound high school and college students goofing off and posting about their antics seems to be over, for better or worse, and universities are certainly trying to end it once the young women and men step onto campus. Perhaps play, goofing off, and silliness do not belong on social media at all, yet the systematic "cleaning up" of anything that doesn't belong on the highlight reel is contributing to a success- and achievement-driven culture of young adults who know only extremes—how to be single-mindedly focused on the task of becoming job-ready, often by doing a whole host of activities that prove they are superachieving.

That young adults are finding outlets for play in Snapchat and Yik Yak is helpful, even though both platforms feature their share of problematic and risky behavior. But the fact that these are fast becoming the *only* outlets for such behavior—that anything that isn't part of the highlight real must disappear or be anonymous—is disconcerting.

7. THE VIRTUE OF UNPLUGGING: THE CAPACITY TO SIT, THINK, AND BE ALONE

Because social media and smartphones are exacerbating a go, go, go, high-achieving, success-driven culture, it has never been more important to consider our ability to stop and be still. So many students expressed both their belief that because of smartphones there exists an expectation that they be available and "on" 24/7 and a concern that because of smartphones and the near-constant compulsion to check their social media apps, people do not know how to be alone anymore. Our capacity to be patient and our attention spans are becoming casualties of the ubiquity of smartphones in our lives.

Yet our capacity to be alone, to slow down, to be still, to rest, to do *nothing at all* is essential for health and well-being, for decision-making, for self-understanding, happiness, peace, and relaxation, to catch those random moments of beauty that the young man spoke about in my class. Young adults need the time and space to get to know themselves, to reflect on who they are and what they want from life, from their communities, from their relationships. Security, self-esteem, and personal well-being are connected to our capacity to be alone and to be comfortable with ourselves and our thoughts. Just as we need the time and the right to be playful and fun, we need the time and the right to ponder, to watch and to listen, to pay attention to the world around us.

We must ask ourselves: How can we do less? How can we learn to be alone with our thoughts? How can we become more comfortable with boredom and stillness? How can we overcome, at least sometimes, the compulsion to grab for our phones every time we are waiting in line or walking alone across campus, or when there is a lull in conversation? We need to think about ways to structure our lives so we don't have to be available all the time, so that we can have designated times when we are unplugged. It is essential to our happiness and well-being.

8. THE VIRTUE OF QUITTING: WHY GIVING UP CAN BE EMPOWERING

The notion that we should never quit, that we should try, try, and try again, is a popular ideal in American society, but I don't think we should hold to it when it comes to social media and our smartphones. Knowing when to stop, when to set something aside, how to decide when something isn't working and it's time to move on—maybe not forever, but at least for a while—is a good skill to have. It's never been more essential than it is now.

The stories of students who quit various platforms on social media either temporarily or permanently should force us to consider the importance of knowing why and when to do this, as well as the benefit of being able to let go in order to choose a healthier path forward. They are doing their best to figure out a positive relationship to social media without much guidance, largely because we are all struggling with the very same things ourselves. The students who quit not only are standing up for themselves but also serve as examples of the importance of drawing boundaries, of acknowledging limits, of being able to step back and step away, and the benefits of doing so. Even a temporary respite can help empower us to gain a new perspective, critical distance, emotional repair, and a kind of freedom in living apart as opposed to desperately treading water or remaining in a place that makes us feel like we are out of control.

We do not have to be powerless in the face of new technologies, in the face of social media and our devices. Quitting either permanently or temporarily, on a grand scale or a much smaller one, may be one of the healthiest steps we can take. The degree to which one steps away is personal and individual, yet the wisdom of doing so is something I believe that all of us need to consider.

THE GOLDEN RULE OF SOCIAL MEDIA FOR OUR
NEW AND VIRTUAL WORLD

By considering these eight virtues in relation to our online lives and teaching our young people to do the same, I believe we can at least *begin* to develop a healthier relationship to social media. It may make a lot of sense not to bare our souls and our deepest feelings on Facebook. But if we decide this is what we believe, then perhaps we will be required to change our overall relationship to social media, lest we risk our souls and our emotional lives in the process. If social media truly requires such filters to maintain a healthy public and professional life, then perhaps we need to renegotiate our use of it.

This is not to say that we must develop specific guidelines for all individuals on how much time, checking, and posting is healthy or normal. As Aristotle suggests, each person's virtuous "mean" will likely be different. But I do believe we need to develop a framework for thinking about our needs in relation to our usage, so as we raise and teach our children, and consider this in our own lives, we can come to some decisions about how we are going to use (and not be used by) these platforms.

All of which is to say that, if there is one overarching rule that can be said to encompass these virtues and help us better navigate the shifting world of social media, it is another one borrowed from the ancient Greeks: know thyself. Know your limits. Know what you can handle and what you can't. Know what upsets you and what makes you happy. Know what you value and what you don't. And, most important, know who you are offline so that what you see online doesn't give you a distorted self-image. By knowing ourselves, we can take back the power social media has over us, if we do indeed determine that it has taken too much power away.

But knowing ourselves takes time and work and intention.

Taking the time and doing this work is *essential.* We need to carve out the time and make this investment at home and in our classrooms, be they in high schools or on college campuses.

Over and over again, students discussed how, ideally, social media and smartphones should be "tools" that serve our self-expression, our sense of purpose, and our relationships. I think we need to reckon with the possibility that if social media and smartphones *are* to be tools, many of us may need to change our habits.

One of the best places to negotiate these changes, to critically analyze their need and significance, and to bring all the resources of a multitude of disciplines to bear on this endeavor is within the classroom at our high schools, colleges, and universities. Critical thinking is what we are meant to do in the classroom and navigating this new and virtual world requires critical thinking on a scale we've yet to devote to it. Universities in particular are places where some of the greatest minds alive have gathered to think through the most serious issues and most difficult challenges and changes we face in our culture and the wider world. We likewise draw the most gifted minds of the next generation to our campuses, and our job is to challenge those minds to engage with the world as it is, not simply as it has existed in the past.

College is supposed to be a place to ask the biggest most important questions a person can ask. We need to be asking big questions about social media and smartphones, too, outside the classroom of course, but *inside* the classroom, too. We cannot abandon our students and the next generation to navigate the stormy waters of social media and smartphones on their own. It is the *responsibility* of university administration and faculty to take seriously the challenges we face because of the dramatic changes social media and smartphones have brought to our world. To ignore this or turn away with our noses in the air because social media is "frivolous" and so very contemporary is to fail the young people who populate our campuses and lives, and to do so is a failure to see the intellectual and academic implications that social media and new technologies touch on in every one of the humanities and social sciences, and in the most high-minded of ways.

Today's college students are truly pioneers: they are among the first to not only grow up on social media but also have their lives shaped by it. They are part of the first generation that has had to react to others' online mistakes and modify their own behavior accordingly. The themes their stories raise are incredibly provocative. Given that they emerge out of the sphere of higher education, I hope they will become a reason for many of us to take on the responsibility of thinking critically and reflectively, not just reactively, about the ways in which social media is changing how we think about ourselves and interact with the world—in both its real and virtual dimensions. And I hope that with time and work and intention devoted to rethinking the advice we offer the young, we will be able to engage in a new, productive, and *intergenerational* conversation about the lives we are living and creating online—one that will empower us to move beyond the happiness effect and the professionalization of all that we do and are and feel.

Taking Control of Our Smartphones: How Student Affairs Professionals, Faculty, and Parents Can Help Young Adults Feel Empowered with Respect to Social Media and Their Devices

Having interviewed nearly two hundred students, conducted a survey of several hundred more, and had hundreds of informal conversations with students, faculty, administrators, and parents over the years, I have come to recognize two things especially clearly. First, smartphones and social media have essentially taken over young people's lives. Second, young people don't feel they are well-equipped to handle this sea change. Those of us who work with them need to tackle these challenges head-on to help them acquire the tools to use social media in a way that makes them feel empowered rather than oppressed. This takes work.

It needs to happen in the home.

It needs to happen in the university classroom and in our schools.

It needs to happen in our wider communities and in our social lives.

The work ahead of us is necessary and urgent. Social media and smartphones have transformed our lives at such blinding speed that we have a lot of catching up to do.

Academics can be slow to react to these kinds of rapid cultural changes. Scholars often consider anything that smacks of trendiness or pop culture to be unworthy of study. This means that in the halls and classrooms of our colleges and universities—the very places where young adults go to become good critical thinkers and citizens of the world—faculty will often regard the most potent influences on our daily lives as irrelevant to the lofty art of traditional scholarly inquiry.

I mention this because, as a scholar myself, I know all too well that as soon as faculty see the word "practical," many of them tune out. But attending to the ways in which social media is changing our lives and our world is not only the work of the Student Affairs folks and of parents. It is *also* the work of faculty. It is in the classroom that we teach our young people to reflect critically on the world—or at least, we should.

What follows is some practical advice for parents, for those of us in mentoring and programming roles, and, yes, for faculty, too.

SMARTPHONES ARE LIKE LITTLE DICTATORS (BUT THEY DON'T HAVE TO BE!)

College students worry that they do not know how to be alone, how to be bored, how to sit with their own thoughts, or how to handle uncomfortable social situations or the silences that sometimes happen in conversations. They *do* know that all these things make them anxious, and they resort to the compulsive checking of their phones and their various apps to avoid having to deal with this reality. Phones provide them a constant and necessary respite from such a fate. Students both love and hate their devices because of this.

I think often about how powerless the students I interviewed felt in the face of their smartphones, their inability to unplug, to not check their apps, to take much-needed breaks from the 24/7 news cycle of social media. And then I think of that small percentage of students in the online survey who felt they didn't need to have periods of intentional unplugging because they had this thing called willpower—and I recall the superiority I detected in their tones as I read what they wrote. Those students know that they are special and unusual in a culture where most people do not have the willpower to unplug.

Technology should be a tool for expression and connection, not a dictator of our emotions and in our social lives. As one student interviewee contended, we should use it, and not the other way around.

The thing we have to learn is *how.*

We need a dramatic change in our relationship to social media and its delivery devices. We need to shift our culture so that we have a healthy relationship to new technologies. I think both the home and the college campus are ideal testing grounds for such a shift, and that there are simple things that can help us can begin to change our relationship to social media and smartphones, and the way they affect our sense of self and our social lives.

Number one on this agenda: we need to help each other to unplug in the home, in the classroom, on the practice field, and on campus.

And let's be real about willpower. It has limits.

Faculty have begun to ban all technological devices from use in their classrooms because their students simply cannot resist the temptation to check their social media accounts and surf the Web during class. We're returning to the age of pen and paper because if we want our students to learn and pay attention, there isn't much other choice.

I applaud such a decision. It's realistic.

Simply hoping for "willpower" among young adults, or even demanding it outright, isn't a credible proposition. We are all too dependent (even codependent) on our devices. Perhaps only when we begin to experience regular pockets of required unplugging will we begin to develop the skills necessary to resist our devices and become more accustomed to not having our devices within reach.

I do think we need to fully take in this reality: that in order to create environments where listening, learning, and engaged participation can take place, many faculty are *resorting* to banning all technology (even if this is the honest, practical fix to a lack of willpower and the availiblity of Wi-Fi throughout campus). What faculty (and students trying to study) are facing, is the fact that listening, learning, and engaged participation is, more often than not, stymied in the presence of our devices and with the availability of Wi-Fi. Surely, in some instances and for certain disciplines those same devices become necessary tools. But, as a culture, and as institutions of education, we need to take a long hard look at the ways that social media, Wi-Fi, and our devices are *disrupting* the educational process as opposed to facilitating it.

For parents: Why not have a basket readily available for certain times of the day when *everybody* surrenders their phones? During dinner, while doing homework, even while sitting and watching television together in the living room. It's a simple thing to do that empowers everyone to let go for a spell and be in each other's presence or truly sit down to concentrate on something for a sustained time. And certainly, during family outings and vacations, consider requiring "unplugging" for all, or at least part, of the time.

For faculty: ban all devices in the classroom if you believe this will help everyone focus and learn. Just do it. This may not be a perfect response, and your students may protest at first, but eventually the relief and the rewards of the technology-free classroom will grow on them. Obviously, this is not an imperative for everyone, but it will certainly not hurt a college student's education to have a mix of classroom experiences, at least some of which take place entirely offline.

For coaches, mentors, and Student Affairs professionals: hold device-free practices and programming. It's a simple shift, but it will make an enormous difference in the experiences of the students who participate. The more "unplugged programming," the better. Imagine what it would be like if there were opportunities and programs during first-year orientation where no smartphones were allowed. Imagine how different it would be among the first-year students if they couldn't resort to using their devices. They would talk more, meet more new people, and have to learn the valuable skill of not simply grabbing their phone if they get nervous or if a silence occurs within a group.

To everyone: don't be afraid to take this step. If I've learned anything from the students who've participated in this study, it's this: even if it's difficult at first to set aside those devices, eventually the payoff—attention-wise, learning-wise, and socially—is enormous. It will help people to slow down and take a step back. It will create sustained and regular experiences of being present, being "alone," and having thinking time, and it will give young adults the opportunity to learn how to do all those things they are becoming so afraid of, like just being still. Students are looking for help putting down their phones. Give them plenty of reasons to do so.

Regular social media sabbaths will help students begin to feel like checking their phones and social media apps is a choice rather than a compulsion.

CREATING WI-FI-FREE OASES IN OUR LIVES AND COMMUNITIES

I think we were all too hasty in wiring everything in our homes and on campuses and even in our parks without thinking about the possible consequences. Yes, the ease of connection is great! We can get online everywhere we go, even on vacation! But this has also become a terrible burden. Most young people—*most people*—don't have the willpower to disconnect.

Just as we need to empower each other to disconnect for regular periods by, at times, banning devices in the classroom and, at times, during sports practices, in the home, and even for programming on campus, we need permanent, identifiable Wi-Fi-free spaces in our communities and on campus.

College students love those dead spots. It's where they go to study. It's where they'd *like* to socialize, at least sometimes, if the option were available. It's where they'd be able to sit and think and learn to be still and alone with their thoughts. I heard too often from students about a tucked-away corner in the library where the Wi-Fi doesn't reach and how absolutely everybody went there to try to get work done because it was the only place on campus where they could actually concentrate.

Yet the Wi-Fi-free spaces the students find now are accidental ones. We need to become intentional about creating those spaces. We need to offer them up to everyone in our community for reflection, study, thinking, teaching, and socializing. It is a mistake to wire absolutely everything. It creates a situation in which everyone is constantly battling to *not* pick up their devices instead of being in the moment.

So: What about a Wi-Fi-free cafeteria (especially on those big campuses with lots of cafeteria options)? A Wi-Fi-free lounge in the residence halls? A Wi-Fi-free coffee shop or cafe? A Wi-Fi-free spiritual space?

Students will jam those spaces. They'll go to them to study, to socialize, to eat, and to have blissfully uninterrupted conversations. Students spoke so much about the joys of not having Wi-Fi on vacation, on service and camping trips, and during study-abroad experiences. We need to create identifiable public spaces where students (and all of us) can take "mini-vacations" from being online.

Likewise, what if colleges were to offer their faculty a selection of classrooms where Wi-Fi is simply not available? Then a faculty member could pick and choose what kind of classroom experience they would prefer for teaching and learning ahead of time, rather than have to go through the battle of banning devices on the first day and ensuring the students don't succumb to temptation during class on subsequent days.

RETHINKING COLLEGE APPLICATION
AND CAREER CENTER ADVICE
ON SOCIAL MEDIA

We need to rethink the advice we are giving teens and young adults about how to polish and clean up their online profiles. Or, we at least need to better understand the consequences of this advice. We need to ask ourselves: Is it ethical that we check the accounts and profiles of college applicants, of potential employees, of our athletes and residence hall advisers? What is it doing to the emotional health and well-being of young adults to know they are being evaluated and judged on platforms that originated as spaces for socializing and self-expression? Is the cost of doing this too high for the young people we care about?

We have turned social media into new stages for performance, new proving grounds for success, new spaces to showcase achievement, new places where young adults must work to meet the expectations of people they hope to please.

I believe that we (those of us in the position to judge, including well-meaning parents) are contributing to the extremes students experience—of over-achieving and perfection, on the one hand, and vile behavior (as on the anonymous Yik Yak), on the other. I do believe that a message that started with good intentions has had significant negative effects on the social, emotional, and intellectual well-being of young adults. While the consequences may be unintentional, this doesn't mean we shouldn't confront them and change course.

We may simply need to stop using social media as a place to measure, judge, and evaluate. We may simply need to acknowledge that none of us is perfect, and that potential employers and admissions officers don't belong on applicants' social media accounts. And we can't let the excuse that we are ensuring our child can keep up with the competition let us off the hook in confronting this issue. I can hear people now: "But I want my child to get into a good college, so she *has* to toe the line like everyone else!" And, "We want *our* students to impress future employers, so regular 'cleanups' of their social media accounts are essential!"

If these questions are inevitable, we must follow them up with others: Do I want my child to feel the crushing pressure of performing constantly for others and being inauthentic online? Do we want our students to learn to navigate who they are in public honestly, or do we hope they all work to become polished "name brands" while they walk (and eventually leave) our halls?

Do we want the young adults we care about to learn that the *appearance* of happiness is more important than actually being happy?

I would guess that the answers to this second set of questions (for anyone who cares about the health and emotional well-being of their children and students) will generally amount to no—no, we don't want these results for young adults and students. This is not to say that we go straight back to the days of college students posting their keg stand photos on Facebook (among other inadvisable things) and I do remember those days, since my Facebook account dates back to the earliest era of the platform when each individual college had its own Facebook, as yet unconnected to other colleges, and where my students posted images and updates that I wished I could unsee. But we need to take a long hard look at how our well-meaning advice is contributing to the advent and popularity (even the necessity) of sites like Yik Yak that allow young adults to do all the things (and far far worse) they feel they no longer can on sites attached to their names. Yik Yak is a bit of a monster, and I do believe that we (the well-meaning adults) need to consider whether our advice to young adults about social media has inadvertently helped to create it.

THE IMPORTANCE OF INTELLECTUAL REFLECTION ABOUT SOCIAL MEDIA INSIDE THE CLASSROOM

Very few students who participated in this research could identify classes on their campus that asked them to reflect on social media and smartphones and how they are changing our lives, our sense of self, and our world. The discussions that do happen seem to occur in the business school, where students learn about branding, advertising, and how to use social media for their careers. The conversation centers on social media as a marketing tool and stops there. The other department on campus where students might find such discussion is in courses on media and journalism, but again, these are courses about how to use social media to expand one's audience or about the importance of social media to the dissemination of information.

If the topic of social media came up for discussion outside of these two spheres, it was only in cursory ways, such as telling students not to use smartphones in class, or that they should use Facebook to set up group meetings or even set up a class Facebook page for communication and posting.

Thus, we seem to be either relegating social media and smartphones to their professional uses (in business or media/communications) or merely considering them as tools. Until now, we seem to be following that "professionalization of Facebook," Career Center–type model of drawing on social media and smartphones only when they touch directly on professionally related concerns.

At only one school did I hear of a class where students were asked to consider how social media was affecting our self-understanding and our relationships. This discussion is the one that is needed the most.

As I said in earlier chapters, the capacity to think critically and philosophically about the role and influence of social media in our lives seems to empower young adults to make better decisions in their use of it, helping them to understand what they see on it, as well as how their own participation influences their sense of self and real-life relationships. We need to actively empower our students to do rigorous analysis of how social media and our devices are influencing all those big questions we ask (or should ask) in college: about how they are influencing our emotional, personal, familial, and relational lives; how we experience childhood; how we make choices; our sense of well-being and self-esteem; and, certainly, our happiness. One excellent place for this is in the classroom (whether that classroom is on a university campus or in a high school).

This does not mean that we need to offer entire courses devoted to such topics (though we do need at least some course offerings of this nature), but we must ask direct questions about these subjects when we are teaching philosophy, psychology, theology, politics, sociology, and literature, which invite students to make connections between what they are studying and how they are living their lives.

TEACHING STILLNESS, MEDITATION, SLOWING DOWN, AND JUST BEING

In our efforts to wire everything, "connect" everything, update and upgrade classrooms, homes, and residence halls, we've made it incredibly difficult to just sit still and be. In any and all moments of stillness, of doing nothing, everyone can now reach for a device that will distract from the quiet of being offline. Even going for a run is now interrupted by Fitbits that monitor our heart rates and our number of steps (among many other things)—which is nice in theory but is yet another distraction from, say, just paying attention to the color of the leaves or letting the mind wander here and there. Now we can monitor and count and play with yet another device that delivers information and gives us specific objectives and goals, something to attend to other than our own thinking about life and the world, ourselves and the people we love, or the simple rhythms of our body as it moves.

When my own students and the student participants in this research discuss their worries about their inability to be alone, about how they no longer know how to endure silences in the presence of others, of how they fear listening to their thoughts, of how their smartphones don't allow them to ever slow down, I often think about the overwhelming interest they also show in spirituality and spiritual practices—as demonstrated in my work for *Sex and the Soul* as well as the research of other scholars, most notably Christian Smith. What's wonderful about such practices is that they are frameworks for doing exactly the things that seem to be slipping away from us: slowing down, being still, listening to our thoughts (or letting them go altogether). Spiritual practices and religious rituals tend to involve stepping away from the chaos and distraction of everyday life in order to center ourselves in another way, in another place, on the inner workings of our hearts, our souls, our minds. Here is where I also think about the young woman who spoke of going to church on Sundays merely because it was one hour away from her smartphone and how she needed this respite—and needed a reason to take it.

Whatever tools we can find to help us carve out this kind of time in the home, in our schools, in the classroom, and in the residence halls is useful and important. But this is also where Campus Ministry programming can do a lot of good and be particularly useful.

Spiritual practices, retreats, and religious services not only are opportunities to unplug for short or extended periods of time, or even regular periods of much-needed space for quiet, stillness, and thinking, but they also offer frameworks for conversation, reflection, and contemplation of the ways that social media and smartphones are affecting and changing our lives. Just as the classroom can—and must—become a sphere for critical reflection on these subjects, the programming, rituals, and practices associated with religious traditions and Campus Ministry can become opportunities for us to take a step back and wrestle with how we are handling (or not handling) these new influences on our lives, our actions, our relationships, and our communities.

This can happen on any campus, of course, since most universities (including nonreligiously affiliated ones) have resources and centers from a range of religious traditions available to their students. But this can also happen at high schools (at least private and religiously affiliated ones) and at home within families that either practice a religious tradition or take up some form of alternative or individual spiritual practice.

TAKE THE UNPLUGGED CHALLENGE!

For those on campus who are in charge of outside-the-classroom programming such as the First-Year Experience programming that happens in August when students arrive at college, and the subsequent year-round programming sponsored by the residence halls and Student Activities, consider the ways in which this programming can be used to encourage students to unplug regularly.

Everyone knows that the promise of free food draws students to events—why not tie that free food to the requirement that students leave their smartphones in their rooms for an hour of device-free socializing over dinner? (Actually advertise this way: Pizza, Unplugged! In the 3rd floor lounge at 8!) Or hold a residence hall–wide contest during exams where students track their smartphone-free/unplugged hours, and whoever clocks the most concentrated study hours (or the top three people) wins a gift card (or something like this).

Much of the outside-the-classroom programming that happens on campus happens where students live, and providing regular, fun opportunities that empower students to set aside their devices and socialize and study without them both help the students relax and concentrate (if they are studying), and help them get used to leaving aside their devices for events on occasion.

LET'S TALK ABOUT OUR RELATIONSHIP TO SOCIAL MEDIA (ON CAMPUS AND AT HOME)

Likewise, just as young adults need opportunities (and the impetus) to unplug regularly and be in Wi-Fi–free spaces on occasion, one of the amazing things about being at a university (and at many high schools for that matter) involves the incredible range of resources available to students devoted to their health and well-being—and we need to actively put those resources to use toward the end of helping young adults talk through how they feel about social media, how it affects their relationships with others, their sense of self, and self-esteem, as well as the phenomenon of comparing ourselves to others that social media can exacerbate (at times) to an extreme degree among its users.

The Counseling Center on campus (as one example) could openly offer sessions for students struggling with some of the above issues (and actually advertise on campus that the Counseling Center is available for this). On a less formal (but no less important) level, R.A.s could make themselves available for students to come talk about how social media is making them feel, and professional staff members could add a social media component to student staff training, as well as provide the R.A.s with referrals for residents on campus who really need to talk about this subject beyond what an R.A. can provide. Openly advertising to students where they can find resources to think through their relationship to social media (and especially some of the trauma students are experiencing now because of what happens on sites like Yik Yak) needs to become an intentional and structured part of what we do within Student Affairs.

Just as I advocated in *Sex and the Soul* that parents of high school students do a pre-college sex and relationships talk, an open dialogue at home about the psychological and self-esteem side of social media is just as important (and I would dare to say, far more so) as the warnings parents give about what *not* to post on social media. We need to get used to talking about how social media is changing our lives in general with each other and with our children, and not only when it involves image-curating (as when someone is applying to college) or extreme circumstances such as cyberbullying.

POWER AS OPPOSED TO POWERLESSNESS

The more conscious and critical and open we can become in relation to social media and smartphones, the more practiced (and the less compelled) we become in relation to our usage of these platforms and our devices, the healthier our relationship will be to all of it.

As I was conducting this research, I began to think of the Mark Zuckerbergs of our world as the new gatekeepers in our lives. Their creations permeate everything we do, how we think about everything in our lives, what we do in our free time and work time, how we sleep (or are unable to sleep). And each little "update" they enacted over the course of the time I was doing this research wreaked havoc on all the students I was interviewing. It stressed them out, affecting and even destroying some of their relationships in the process.

Take the update that now tells Facebook users whether and exactly when someone has read a message. Or the one on Snapchat that suddenly ranked your friends in order of importance based on how often you "snapped" them. These "slight" changes had huge emotional and personal consequences for the students. They affected when (or whether) they opened messages or communicated with their friends and significant others. They felt toyed with and betrayed by the people behind Facebook and Snapchat.

Right now, we are still operating in a place where social media and our devices are dominating us and our behaviors, far more so they we are simply using them as tools to our own delight and benefit. The students I spoke with *like* social media and smartphones when they are useful tools. They stop liking them when they become forces wreaking havoc on their psyches, their ability to conduct their lives and get things done, and their relationships. Social media and smartphones are not going away. We need to work to get to a place where we know ourselves, our weaknesses and strengths, well enough that these platforms and devices lose the power to upend everything, and so that they remain the useful tools students would like them to be.

As Aristotle might tell us: with practice, we'll get there.

But practice is essential.

Acknowledgments

First and foremost, I want to thank Christian Smith for being a wonderful colleague and friend. When I was first thinking of doing this project, I asked Chris if he would help me talk through whether it was a good idea, and if so, how I might set it up and get good, solid data. An initial phone call lead to a full day of brainstorming in Chicago, and an excitement and enthusiasm about this project from Chris that tipped me from *I think maybe I should do this?* to *I'm definitely going to do this!* I am truly grateful for the ongoing conversations about all of our work, especially this study, and for his faith in me as a thinker and scholar.

I am also incredibly grateful to everyone at the Center for the Study of Religion and Society at the University of Notre Dame, most especially Rae Hoffman, Sara Skiles, all the transcribers, and Nicolette Manglos-Weber, for her help going through the IRB process. The CSRS, the University of Notre Dame, and the Lilly Endowment, Inc., together provided the funding to make this project happen, and I am grateful for their investment.

To the thirteen colleges and universities that participated in this study (who shall remain anonymous): I can't name you here, but I thank everyone on campus who worked to make this project happen, responded to all my very specific requests and needs, helped generate randomly sampled lists of potential interviewees, found interview rooms, provided on-campus support during my visit, and helped administer the online survey. Thank you for your openness and willingness to host this study at your school. It was no small endeavor.

I want to thank Oxford University Press and everyone who worked to make the book for this study happen, and for their investment in this project and in me as an author. I especially want to thank Sarah Russo, Marcela Maxfield, Ryan Cury, and Maya Bringe (who listened to all my last-minute worries and dealt with all of my last-minute corrections). And of course, I want to give my sincerest thanks to Theo Calderara, my editor, who is truly amazing and pushes me to be the best writer and thinker I can be. I am forever grateful for your incredible dedication, your astute feedback, and your amazing friendship. Also, thank you for making that deadline with me!

My friends have also been patient and thoughtful conversation partners for this study and the book. It was fun to be able to talk about this project with all of you and I appreciate your willingness to share your own stories and struggles and joys with social media, as well as share your opinions and thoughts about the findings. I won't name all of you, but you know who you are.

I am forever and always grateful to my longtime agent and friend Miriam Altshuler, whose enthusiasm and excitement for my work makes me feel lucky. Your ongoing support and dedication to my career as both a scholar and a writer means the world to me.

To my husband, Daniel, who endured all of my travel for this project, as well as participated in uncountable conversations about social media, my findings, my questions, providing feedback, opinions, and questions of his own: you were there from the very beginning of this study and I am grateful for all the thought you gave to my work. As well as the many evenings of wine after long, long days of writing!

Most of all: to all the students who participated in this study as interviewees and in the online survey. I am grateful to have met so many of you in person, and to be the recipient of your fascinating stories, opinions, and willingness to think hard about all of my questions. You are the heart of this project. Your openness to share your experiences will be the inspiration for so much conversation and thoughtful reflection among your college peers and far beyond. Thank you.

APPENDIX

METHODOLOGY

BACKGROUND, MOTIVATIONS, AND GROUNDWORK
FOR THIS STUDY

The research on which this book is based, (originally titled: "A Study of How Social Media and New Technologies Are Affecting Identity Formation, Meaning-Making, and Happiness among College Students"), was inspired by discussions over the last several years with college students all over the United States regarding their questions and concerns about social media and how social media is affecting their lives. These conversations arose on behalf of previous research related to my study that was eventually published as *Sex and the Soul: Juggling Sexualiy, Spirituality, Romance, and Religion on America's College Campuses.* College students, therefore, would most often raise the subject of social media in the context of discussing romance, sex, and hookup culture on campus, and typically the conversation would be about how social media is affecting the way "people get together" and "hook up" on campus. More recently, I noticed how those conversations expanded beyond relationships to include general thoughts and worries about how social media is affecting, as expressed by students, "how I understand myself" and also "my broader place in this world." Students began to mention concerns about feeling split in two—the real and the virtual—as well as their worries about how social media was going to affect their futures.

Initially, though I enthusiastically participated in the conversations with students about social media, I had no intention of doing any formal research on the subject. It wasn't until the number and frequency of students who mentioned social media as a major change agent in their lives became so great and so common—and their concerns *so complicated*—that I realized I wanted to investigate social media and how it's affecting college students and campus life in the context of formal research.

The motivation for taking up this project is nearly identical to my motivation for taking up the research that became *Sex and the Soul*. That study, too, arose out of conversations I'd had with undergraduates (but at a single institution as opposed to a range of them) who were concerned about hookup culture on campus. What interested me most, in that case, was how my students claimed they would continue to engage in behavior (hooking up) that they generally did not enjoy and that often made them feel unhappy, even depressed, yet nobody (no adults/professors/administration) was really talking to them about it or providing productive forums for them to reckon with what they felt was pervasive and destructive behavior on campus. If they'd so intensely identified *drinking* (as opposed to hooking up) as both the problem and the absent conversation on campus, I would have investigated that instead. My interest, there, was in trying to understand the following: Was this also a problem for students at other college campuses? If so, what was the scope of the issue? Finally, how could we open up forums for conversations on this subject, both inside and outside the classroom, where students could finally discuss their feelings about it not just with each other but also with professors and administration on campus?

I am not a big social media person myself, though the subject interests me generally. However, as a professor, a former Student Affairs professional, and a writer of books for young adults, I am very interested in any subject that deeply affects and transforms their lives. As with hookup culture, social media became such an obvious subject in need of investigation that eventually I could not turn away from it. What's more, young adults (college students in this case) are struggling to understand how to navigate social media—specifically its pervasiveness—productively in their lives. They are curious to know how their peers feel about it, whether their peers are also struggling and in what ways, and they definitely hope for new forums (especially academic forums) where they can discuss what is going on with social media. My hope in conducting this research was to gather a sense from a diverse range of college students at a variety of campus cultures about what, simply, is going on with social media in their lives, in order to use this research to open up new avenues for conversation that could occur both inside and outside the classroom, and engage students, professors, and administration together discussing the issues. One other hope I have for this research is to reach parents who have children of all ages and to provide them with topics for discussion and reflection within their families.

Social media affects all of us at every age, so when I set out to do this study, I believed that it could provide useful resources for conversation beyond the college campus as well. Once I committed to researching this topic, I decided I would model the scope and collection of data after my earlier study on sexuality and spirituality, which included gathering online surveys and conducting in-depth, one-on-one interviews.

At the outset, I enlisted the help of my colleague Dr. Christian Smith, the director of the Center for the Study of Religion and Society at the University of Notre Dame, and the primary investigator of numerous national studies, to talk through the whys, the hows, and the methodology for the data-gathering portion of this research. In turn, Dr. Smith offered his center and the resources of his associated graduate program in sociology at Notre Dame to serve as a hub for conducting the study. Prior to gathering any data, I attended several planning meetings with Dr. Smith and also several meetings with his graduate students to discuss interview questions and interview methodology.

During these discussions I identified gathering qualitative data as my main interest for this project, despite biases by many social scientists who tend to prize the quantitative over the qualitative. The qualitative data I gathered for *Sex and the Soul* proved the most useful to my research, and with enough qualitative answers about a particular subject in an online survey, quantitative data can also be generated. My background in gender studies helped me to prioritize qualitative data for my earlier research, as did the wealth of essays on feminist methodologies for qualitative research in Deborah L. Tolman and Mary Brydon-Miller's book *From Subjects to Subjectivities: A Handbook of Interpretive and Participatory Methods*.[1] Dr. Smith helped greatly in my current decision-making process to focus on gathering qualitative data, given his range of experience with overseeing national studies and working with various types of data, and to develop a simple way to randomly sample potential candidates at each participating institution.[2]

The interview questions themselves cover a wide range of issues related to social media and were developed to try to address the questions I was hearing from students during lecture visits for *Sex and the Soul*. The online survey was developed after the interview process was completed and included a smaller series of only open-ended questions; those questions were selected based on their salience with students during the interview process.

With the help of both Dr. Smith and Dr. Nicolette Manglos-Weber, I went through the University of Notre Dame's institutional review board (IRB) for both the interview and the online survey dimensions of this study. The Notre Dame IRB approached my application with the understanding that, in theory, the board would be providing approval on behalf of all participating institutions. I also subsequently went through the IRB approval process at all thirteen schools that participated in my study, but Notre Dame's stamps of approval helped to expedite this process.

INTERVIEW METHODOLOGY AND PARTICIPATION: SAMPLING AND ADMINISTRATION

The student interviews are the primary and most important sources collected in this study. To obtain the strongest pool of interview candidates, I employed a random sampling methodology (more on this later).

The interviews I conducted were semistructured and lasted anywhere from thirty to ninety minutes. With every participant I asked about the same series of topics and questions. However, every interview was unique because the students raised different themes and issues; when a student made a comment that was particularly intriguing, or that he or she wished to go into in more depth, I reserved the right to ask additional questions related to that topic. In some instances, this semistructured approach led me to add questions to subsequent interviews, since several participants raised important issues that were not addressed in the basic questionnaire. I did my best to ensure that every interview touched on every main theme and question overall, even if that meant the interview ran very long. All students were informed before the interview started that if they wanted to skip a question, they had this right, but very few chose to do so. Each student was promised anonymity and the student signed an informed consent document prior to the interview.

One of my primary concerns was spreading the interviews out across a range of institution types. When selecting the schools, I considered factors such as religious/nonreligious affiliation and geographic location, and I tried to reach as diverse a student population as possible in terms of educational background, race, ethnicity, and socioeconomic status. I began my search by approaching personal acquaintances and colleagues at a number of schools, as well as taking advantage of new contacts and opportunities provided by lecture visits, if the inviting institution fit the school type for which I was searching. Finding institutional participants was rather easy because everywhere I inquired, colleges and universities were very interested in data on social media and their student populations, and thus the study intrigued them. My campus contacts ranged from faculty members to Campus Ministry to Student Affairs administrators; these contacts helped me pass through the IRB process at their institution, facilitated access to a random sampling of students for the interview pool, and helped identify a space where the interviews could take place.

In the end I visited thirteen schools for the interviews: three public (one each in the Northeast, the Southeast, and the Midwest); three private-secular (one each in the Southwest, the Northeast, and the Midwest); three Catholic (one each in the Northeast, the Mid-Atlantic, and the Midwest); two evangelical Christian (one in the Southeast, one in the West); and two mainline Protestant (both in the Midwest). All participating institutions were promised anonymity and will remain anonymous.

To identify the pool of potential interview candidates, once a participating school passed through the IRB process, my campus contact would generate a random sample of thirty to forty students, trying to balance for gender. At the largest institutions, my campus contact helped identify a way to sample from a large pool of students in a manageable way. At two such institutions (both of them private-secular), this meant the sample was pulled from the entirety of the school's Honors College. At one of the largest public universities, this meant drawing from the pool of students at one of the university's specialized colleges. At another two of the larger institutions (one evangelical Christian, one private-secular), the sample was drawn from the complete rosters of students living in campus housing. At five of the schools (one Catholic, two mainline Protestant, two public), the random sample came from the combined student lists from a series of classes. At the three remaining schools (two Catholic, one evangelical Christian), the sample was drawn from the entire undergraduate student population.

At each school, I interviewed between eleven and sixteen students, for a total of 184 interviews across the thirteen participating institutions. To identify these students, I took the randomly sampled list of names that included their emails and gender that was given to me by the campus contact, and I emailed an invitation to participate in the interview process (with all necessary information and informed consent form attached) to the first eight women and the first eight men on the list. Then I would do my best to get as many of those students as possible to agree to meet with me when I was on campus—ideally the first sixteen I initially contacted. I did follow up with emails and, if necessary, also had the campus contact follow up with the students I hadn't yet heard from to encourage their participation. The response rate was high overall, though at two of the schools, it proved to be unusually difficult to recruit student participants out of such a small random sample.

In the end, to get 184 student participants for the interviews, I emailed a total of 235 invitations to randomly sampled students across the thirteen participating institutions, for a response rate 78 percent. Not a single student who participated in the interview process volunteered based on interest. Once the invitations went out, some students responded nearly immediately, and I scheduled their interview; I needed to email some other students three or even four times—sending reminders and pleas—to get them to agree to meet with me. At times, I called on my campus contact to follow up with the stragglers on my behalf, and this usually generated a few more responses as well.

Each potential participant was offered a cash incentive, which he or she would receive at the end of the interview in the form of a thirty-dollar prepaid credit card, generated at Notre Dame for this purpose. Out of the fifty-two students who received invitations but did not do interviews, forty-three simply never responded to my attempts to get in touch with them, six actively declined, two canceled at the last minute because they got sick, and one canceled at the last minute because of a lab that got rescheduled during the interview slot, on the day that I was leaving campus.

All participating students had to be at least eighteen or older and enrolled as undergraduates.

The benefit of random sampling is obvious: the pool of interviewees at each school was pulled based on the simple fact of being currently enrolled undergraduate students at the participating institution and *not* because they self-identified as being particularly interested in or active on social media or desirous of discussing it. This method of sampling helps to alleviate this sort of bias among the interview pool. One potential weakness in the interview pool involves the five institutions where the random sample was pulled from a series of classes and professors—in other words, from a smaller population of students on campus, as low as eighty to a hundred—as opposed to the eight other institutions, where the sample pulled was drawn from populations that generally ranged into the high hundreds (at the very least) and even thousands, depending on the school's size. Even at these five schools, however, there was not a single volunteer. All participants received their invitations "out of the blue," so to speak, and then we went from there.

Another weakness I can identify that may have affected the response rate is that email is a tough way to contact students today. Many students check their university email accounts very infrequently, if at all. Some universities try to require students to check their campus emails, but students tend to prefer contact via either text or, perhaps a bit ironically, Facebook. Two of the institutions (one Catholic, one private-secular) assigned me an undergraduate to help me track down the students; these student assistants followed up with the stragglers on Facebook, which helped a good deal in signing them up for interviews. I, however, did not contact any potential interviewees by text or via their social media accounts. In the future, if I was to initiate a new study, I would consider finding alternative ways to contact potential interviewees other than via email.

Overall, the demographic breakdown of the interviewees was as follows:

Gender: 92 women (approx. 50%), 91 men (approx. 50%)
Racial breakdown: 121 white; 22 African American; 12 East Asian;
    5 Southeast Asian/Indian; 1 Middle Eastern; 13 Hispanic; 10
    biracial (4 half black/half white; 2 half Pacific Islander/half

white; 2 half East Asian/half white; 1 half Hispanic/half white; 1 half East Asian/half Native Hawaiian)

Sexual orientation: 181 heterosexual, 1 gay, 1 lesbian, 1 bisexual

School type: public: 42/184 = 23%; private-secular: 44/184 = 24%; Catholic: 45/184 = 24%; evangelical Christian: 27/184 = 15%; mainline Protestant: 25/184 = 14%

The topics for the interview included the following:

Highlights of college
Friendships (general)
Romantic relationships (general)
Describe self (generally)
Describe self (socially)
Meaning/happiness (generally)
Religious background
Describe self (on social media)
Why do you participate/post?
Criteria for posts
Feelings about people's reactions/nonreactions to posts
Expressing emotions on social media
Faith on social media? (if relevant)
Have you ever thought about quitting any accounts?
Bullying
Social media history—first account, what age, etc.
Parents and social media
Selfies
Comparing self to others/FOMO
Success and social media
Social media is necessary/unnecessary today?
Online image
Privacy
Gender
Competition/jealousy on social media
Self-esteem
Relationships/romance/dating and social media
Sexting

Smartphone and related issues
Social media and happiness
Discussions of social media at college with professors/staff?
Anything else?

ONLINE SURVEY METHODOLOGY AND PARTICIPATION: SELECTION AND ADMINISTRATION

The questions for the online survey were drawn directly from the pool of questions used for the interviews, and they were chosen after I finished all the interviews at the participating thirteen institutions. I waited to administer the online survey until after the interview process was finished because the interview process helped me to select only the topics that seemed most important to the students for inclusion in the online interview. I wanted the online interview to be as short and as efficient as possible—especially because it consisted of open-ended essay questions.

All questions were open-ended because—in my opinion—this format allows students to explain why they feel they way they do and why they do what they do, giving us far more information and insight into the landscape and significance (or insignificance) of a subject than questions that merely seek quantitative statistics.

All questions were also optional—save two. The first of these asked students either to check the social media sites to which they belonged or to check that they had none, which would send them to a different set of questions about why they don't participate in social media. The second asked students if they had a smartphone; again, those who checked "no" would be sent to a different set of questions. The decision to make all the essays optional was made both to encourage the ease of participation and to see which topics really interested the students taking the survey enough that they chose to answer them.

All students were required to give informed consent for the survey before they could proceed to the survey questions. All students also had to be eighteen or older to participate and had to be enrolled as an undergraduate at their participating institution.

Aside from being asked to provide basic demographic data, students were asked to answer a gridlike series of questions about how various issues on social media make them feel, before moving on to a series of open-ended essays—all of which were optional. On the main page for essays, students were asked to choose five out of ten topics to write about, before moving on to a series of optional essay questions that asked about relationships and then another series about smartphones.

The following is the list of all optional essay topics students could answer for the online survey, in abbreviated form:

FIRST ESSAY PAGE

- Dos and don'ts of social media
- Social media behavior—then (when you first got on) versus now
- If you could go back, what do you wish you'd known about social media?

SECOND ESSAY PAGE (CHOOSE 5)

- Social media and self-expression
- Most/least favorite thing about social media
- Selfies
- Online image
- Social media as obligation
- Rules/criteria for posting
- Comparing yourself to others
- Gender and social media
- Anonymous sites
- Quitting social media

THIRD ESSAY PAGE

- Apps like Tinder
- Friendships
- Sexting

For students who have a smartphone:
- What do you like about them?
- Does having one mean you are "always available"?
- Do you ever intentionally take a break from it?

For students who do not have a smartphone:
- Why don't you have one?
- How does not having one affect your life/college experience, if at all?

Please note: students who did not participate in any social media were directed to a special essay page designed for this, immediately after they filled out the demographic information. On it were four essay questions that dealt with the following topics:

- Why do you stay off social media?
- Have you always been this way, or did you quit?
- How docs not having it affect your life/college experience (if at all)?
- What do you think of the phenomenon of social media?

Also, all students were given the option at the very end of thc online survey to add anything else they'd like to say, using an essay format.

Unlike for the intervicwees, who were sampled randomly, all students who took the online survey volunteered to do so. My reasons for this were twofold. The first reason was practicality. Convincing randomly selected, nonvolunteer students to participate in a study is incredibly time-consuming and very labor-intensive. The second reason has to do with the potential results: I was interested to see how the results would differ depending on whether the students (1) had been randomly sampled as with the interviews or (2) had volunteered because they were interested in the subject matter. I thought it would be valuable to have both types of participation to be able to compare them.

The obvious weakness of my sampling method for the online survey is exactly this: it's an all-volunteer student pool. But, in my opinion, the fact that the interviews were all randomly sampled and therefore provide a comparison pool against the volunteers not only helps compensate for this weakness but actually helps to strengthen the results of the study overall.

Another thing to note about the online survey is that only nine out of the thirteen institutions where I conducted interviews participated—the three Catholic, the three private-secular, the two evangelical Christian, and one of the public institutions. At two of the public institutions, it proved too difficult to get approval to administer the online survey to students. Many larger universities have restrictions on how frequently they allow their students to be surveyed, and they often prioritize their own surveys over those that come from outsiders such as myself; they do not want their students experiencing "survey fatigue." At one of the mainline Protestant schools, I simply got through to the people who needed to approve the online survey too late for the survey to go out (students were already well into final exams before summer); at the second mainline Protestant school, I never heard a response from my contact there about administering the survey at that institution.

As with the interviews, the students were invited to participate via email by my campus contact at the university. At three schools (two Catholic, one evangelical Christian), the survey invitation was sent out to the entire population on campus; at two schools (one evangelical Christian and one private-secular), the survey was sent out to the entire population of students who lived in undergraduate housing; at two schools (both private-secular), the survey was emailed to all students in the Honors College; at the public university, it was sent to all students at one particular undergraduate college; and at the remaining Catholic school, it was sent out to the population of students taking theological courses during that particular spring semester.

There was no cash incentive to complete the online survey. A total of 884 students across the nine participating schools volunteered to take the survey, which opened in March 2015 and closed in June 2015. Though the response rate was low, the students who did volunteer to take it wrote extensive answers to at least some, if not all, essay questions.

The demographic data among the students who took the online survey break down as follows:

Gender:	Male 26.76%; female 73.19%; agender 0.23% (2); transgender 0.23% (2); other 0.59% (5)
Race:	White/Caucasian 67.56%; Hispanic 7.73%; black 8.55%; East Asian/Asian 7.26%; Native American 0.7% (6); South Asian/Indian 2.11% (18); Middle Eastern 0.82% (7); other 5.27% (45)
Sexual orientation:	Heterosexual 90.21%; gay 3.34%; lesbian 0.95% (8); bisexual 5.49%; transgender 0%
School type:	Public 11.48%; private-secular 35.6%; Catholic 24.71%; Christian, non-Catholic 28%
School year:	First-year 32.32%; sophomore 21.43%; junior 27.4%; senior 18.85%

This book by no means exhausts the qualitative data I collected from all the student participants, whether from the interview or the online survey, and will not be the only product that will result from this national study.

A LAST NOTE ABOUT THE AUTHOR'S PERSONAL LOCATION IN RELATION TO SOCIAL MEDIA

I am not now, nor have I ever been, "a social media person." I had a Facebook profile for many years and then deactivated it because I didn't ever post and, like so many of the students I interviewed, often didn't love what I saw of others when I looked at the newsfeed. There is something about the publicness of it all, in my own life and the way that it exposes the lives of others, that makes me feel stressed. Although I enjoy seeing pictures of my friends' babies, I'd rather see their actual babies. I want to know all about my friends' lives, but I'd rather my friends update me when we meet for coffee or dinner. I tried Twitter for a while and failed miserably at it. I just don't have anything to tweet. I don't take the time to think of what to say. I'm not a multitasker. I still read the newspaper, the kind that gets printed on actual paper. The same goes for books.

I don't have a smartphone either. I never have. I've joked that I am a conscientious objector, but I'm not entirely kidding when I say this to friends and colleagues. I still have a "dumb phone," which can get calls and can text, but I have to push the number buttons multiple times to type out the letters I need, and it takes forever to get out a sentence. When I leave my house, I leave the Internet behind, and I prefer it that way. Like just about everyone I know and the students who participated in my study, I struggle with the compulsion to check email and sometimes find it almost impossible to concentrate and do my work when I could just spend time looking at real estate listings and food and cute cat videos. By choosing not to have a smartphone, I am choosing to free myself of that compulsion when I'm out in the world. It's part of how I manage the ways that new technologies are changing my own life.

So what am I doing conducting research about how social media affects the lives of college students, and how will my own position on it affect what I've written about in this book?

First, even though I'm not currently active on social media, I am young enough to be on the edge of the first generation that it truly changed. Social media permeated my twenties and thirties, and it fascinates me endlessly. I know enough about it to understand both its attractions and its dangers; I can boast enough experience with it to have both enjoyed and struggled with it.

But more important, I believe that my presence or absence online and with respect to owning a smartphone stands apart from who I am as a scholar and professor, a dedicated teacher, and a person whose research has revolved around issues of concern to young adults and college students for well over a decade. I do my best to listen to what college students want to talk about, to hear their concerns, consider their questions, and take their struggles seriously. Social media is so often on their minds, but they feel that the adults in their lives aren't really paying attention to it in the way that they should. So, in my role as a teacher and researcher, I've done my best to attend to what these students are concerned about. I'm most interested in how young adults feel about what they post and do and see online and on their devices, because that is what *they* are most interested to know.

I am sure that some people will see my own relationship to social media and smartphones as a potential weakness of this work, but I hope that most readers and colleagues will trust in my profound concern, deep respect, and love for the students I have met and worked with over the years and will understand how this grounds my research. And while many academics and writers use social media and smartphones as often and as adeptly as the students described in this book, as someone with a bit of distance, I hope I have offered a different sort of wisdom on the stories the students tell and the questions they raise. Social media and smartphones are so pervasive that I believe we need a diversity of voices and perspectives about what they mean—and how they are changing the meaning of our lives. I hope that the research presented here provides a new and thought-provoking window into this subject.

NOTES

Introduction

1. All names are pseudonyms, including those referring to sororities and fraternities, as well as any friends or school-sponsored events mentioned by the students.

2. Throughout this book, I have chosen—out of respect for the students who participated in the interview process, on behalf of consideration for my readers, and to improve the fluidity of the prose—to edit out unnecessary words such as "like," "um," and "oh" and other vocal hesitations that people use as they talk, as long as doing so does not change the meaning of what is said.

3. There are a number of well done book-length studies that explore the millennial generation's relationship to social media and new technologies more generally, however. First among them is danah boyd's *It's Complicated: The Social Lives of Networked Teens* (New Haven: Yale University Press, 2014), which is an effort at translating for parents and others why teens love social media so much, and why it's a useful, identity-building aspect of their lives, as opposed to the destructive, dangerous force that so many people fear it is. Then, in *The App Generation: How Today's Youth Navigate Identity, Intimacy, and Imagination in a Digital World* (New Haven, CT: Yale University Press, 2013), Howard Gardner and Katie Davis also explore the ways in which social media is changing young adults' sense of identity today, though they do so primarily in negative ways.

   Then, a number of more academic, book-length treatments look at social media and how it is shaping (and reshaping) our world overall, regardless of age and generation. In *Social Media: Usage and Impact* (Lanham, MD: Lexington Books, 2012), Hana S. Noor Al-Deen and John Allen Hendricks provide a comprehensive and scholarly analysis of social media, with contributions that examine the implementation and effect of social media in various environments, including educational settings, strategic communication, advertising, public relations, politics, and legal and ethical issues. In the edited collection *The Social Media Reader* (New York: NYU Press, 2012), Michael Mandiberg presents pieces on peer production, copyright politics, and other aspects of contemporary (Web 2.0) Internet culture, including collaboration and sharing and the politics of social media and social networking. In *The Culture of Connectivity: A Critical History of Social Media* (New York: Oxford University Press, 2013), Jose van Dijck describes the rise of social media in the first decade of the twenty-first century up until 2012, providing both a historical and a critical analysis of the emergence of major platforms in the context of a rapidly changing ecosystem of connective media. Looking at five major platforms (Facebook, Twitter, Flickr, YouTube, and Wikipedia), van Dijck notices similar technocultural and socioeconomic ideological principles guiding their development as well as similarities between these platforms' shifting ownership status, governance strategies, and business models. Finally, in *The Social Media Handbook* (New York: Routledge, 2013), Jeremy Hunsinger and Theresa M. Senft present a collection of essays that explore how social media are changing disciplinary understandings of the Internet and our everyday lives. Rather than considering social media in terms of specific technologies, chapters in the book engage topics across a range of research techniques, practices, and theories and address broad topics, including community, gender, fandom, disability, and race.

4. For more information about young adults and narcissism, see Jean Twenge, *Generation Me: Why Today's Young Americans Are More Confident, Assertive, Entitled—and More Miserable Than Ever Before* (New York: Free Press, 2006), and also Jean Twenge and Keith Campbell, *The Narcissism Epidemic: Living in the Age of Entitlement* (New York: Atria Press, 2009). For a comprehensive overview of the Millenial generation (in general), see *The Millennials: Connecting to America's Largest Generation* (Nashville, TN: B&H Publishing Group, 2011), for which Thom S. Rainer and Jess W. Rainer conducted twelve hundred interviews with Millennials in order to better understand them personally, professionally, and spiritually.

5. One of the most divisive topics related to sexting is how it should be handled by authorities and what punishments should be handed out, especially as it relates to sexting by minors. Some legal scholars refer to the images as "self-produced child pornography," and some believe that minors who send sexts should be prosecuted under existing child pornography statutes. Most states, however, have so far taken a somewhat more lenient approach, often allowing juvenile offenders to be charged with a misdemeanor or a lesser offense so they can qualify for diversion programs and have their records expunged. Still others believe sexting by teenagers should be handled by teachers or parents instead of the courts. Perhaps most alarming is the fact that while sexting is considered widespread among teens, most teens are unaware of the consequences of the behavior, such as that sending a sext could potentially be prosecuted as a felony under child pornography laws in some states. For a selection of news stories on teens and sexting see Nathan Koppel and Ashby Jones, "Are 'Sext' Messages a Teenage Felony or Folly?," *Wall Street Journal*, Eastern Edition, August 25, 2010, D1–D2; Jan Hoffman, "States Struggle with Minors' Sexting," *New York Times*, March 27, 2011, Riva Richmond, "Sexting May Place Teens at Legal Risk," *New York Times*, March 26, 2009, gadgetwise.blogs.nytimes.com/2009/03/26/sexting-may-place-teens-at-legal-risk/?_r=0; Maia Szalavitz, "Nearly 1 in 3 Teens Sext, Study Says. Is This Cause for Worry?," *Time*, July 2, 2012, http://healthland.time.com/2012/07/02/nearly-1-in-3-teens-sext-study-says-is-this-cause-for-worry/; and Conor Friedersdorf, "The Moral Panic over Sexting," *Atlantic*, September 2, 2015, http://www.theatlantic.com/politics/archive/2015/09/for-sexting-teens-the-authorities-are-the-biggest-threat/403318/.

6. For readers interested in a detailed treatment of the methodology for both the interview and the online survey process, participant selection, demographics of participation, and so forth, see the methodological appendix at the back of this book.

7. In a nationwide study conducted by the Pew Research Center, survey data gathered in September 2009 showed that 73 percent of online American teens used social networking websites, followed closely by young adults aged eighteen to twenty-nine at 72 percent. Looking at the difference between older and younger teens, 82 percent of those aged fourteen to seventeen used social networking sites, compared

with only 55 percent of those aged twelve to thirteen. Cell phone ownership is nearly ubiquitous among teens and young adults, as is Internet use. See Amanda Lenhart, Kristen Purcell, Aaron Smith, and Kathryn Zickuhr, "Social Media and Mobile Internet Use among Teens and Young Adults," *Pew Internet & American Life Project* (Washington, DC: Pew Research Center, 2010).

8. See Casey Fiesler, "How Missouri Could Demonstrate What's Wonderful about Yik Yak," *Slate.com*, November 12, 2015, http://www.slate.com/articles/technology/future_tense/2015/11/the_university_of_missouri_protests_and_yik_yak.html, for one example of how students at the University of Missouri are using Yik Yak to fight back against such racist commentary from their peers on this same platform.

9. A total of 736 students chose to answer this optional question. Throughout this book, I will provide the raw number of student answers to a particular question in the endnotes each time I refer to survey data so that readers interested in such numbers will have them readily available.

10. According to the 2012 Pew Research survey on social media, 81 percent of teens between twelve and seventeen use social media and 77 percent of them are on Facebook. See Maeve Duggan and Joanna Brenner, "The Demographics of Social Media Users—2012," *Pew Internet & American Life Project* (Washington, DC: Pew Research Center, 2013). The Pew Center also reported that at least 24 percent of teens in this age group go online "nearly constantly." For more information, see Amanda Lenhart, Maeve Duggan, Andrew Perrin, Renee Stepler, Harrison Rainie, and Kim Parker, "Teens, Social Media and Technology Overview" (Washington, DC: Pew Research Center, 2015). Then there is the growing relationship between social media and the college experience itself. In their article "Are Students Really Connected? Predicting College Adjustment from Social Network Usage," *Educational Psychology* 35, no. 7 (2015): 819–834, John Raacke and Jennifer Bonds-Raacke examine the relationships between social network usage and adjustment to college in the academic, social, personal-emotional, and university affiliation domains. The authors' results showed that social network usage was related to college adjustment; specifically, those students who reported

higher rates of social media usage reported lower levels of adjustment to college in all domains.

Chapter 1

1. José Van Dijck, *The Culture of Connectivity: A Critical History of Social Media* (New York: Oxford University Press, 2013), 13. For more on this subject, see also Taina Bucher, "Want to Be on the Top? Algorithmic Power and the Threat of Invisibility on Facebook," *new media & society* 14, no. 7 (2012): 1164–1180.

2. It has been well documented that being on social media can make people feel isolated, and that public sharing of self does not necessarily make a person feel more known or understood. Important to the entire conversation of this book is the work of MIT professor and scholar, Sherry Turkle. In *Alone Together*, for example, Turkle introduces the notion of "I share therefore I am" as a new state of the self, and roots the implications of this and the underlying psychology of it in object relations theory. Turkle is grappling with the idea of a self that is constructed through *showing itself to others*—the construction of self via public sharing—a concept and theory of self that grows ever more important the more sharing everyone does online and in social media. Please see Turkle, *Alone Together: Why We Expect More from Technology and Less from Each Other* (New York: Basic Books, 2011).

3. As with Margaret, Van Dijck's "popularity principle" and its relationship to "likes" applies so clearly here with a student like Rob, too: he is fairly obsessed with accumulating that silent yet visible applause from people pressing that "like" button on his behalf, and he experiences a clear sense of disapproval when people don't press it. He also spends an enormous amount of time during each day trying to figure out how to get more and more of the kind of boost "likes" give him, too.

4. Like the students for this study, the media also thinks a lot about how "likes" are affecting all of us. In her article "How Millennials Use Facebook Now," *Huffington Post*, HuffPost Tech United Kingdom, March 19, Eleanor Moss notes that Millennials know their friends and others online will likely look them up on Facebook, and that by carefully curating a list of "liked" pages, young adults actively create a persona or identity that will make them seem cool, intelligent, funny, or caring, among other favorable characteristics. Additionally,

Moss points out that young adults will often be highly influenced as to whether to follow a page themselves based on how many of their friends have "liked" it; For a more scholarly take, see also "Private Traits and Attributes Are Predictable from Digital Records of Human Behavior," *Proceedings of the National Academy of Sciences* 110, no. 15 (2013): 5802–5805, where Michal Kosinski, David Stillwell, and Thore Graepel found that Facebook "likes" can be used to automatically and accurately predict a range of highly sensitive personal attributes, including sexual orientation, ethnicity, religious and political views, personality traits, intelligence, happiness, use of addictive substances, parental separation, age, and gender. The authors give examples of associations between these various attributes and "likes" and discuss implications for online personalization and privacy.

5. People have long wondered if Facebook would ever add a "dislike" button to go alongside the "like" one, and many have wished for such a feature. Though Facebook continues to resist the "dislike" button, they added a series of five new emojis "reaction buttons" meant to convey: "love," "haha," "wow," "sad," and "angry." For an analysis of these new buttons, see Will Oremus, "Facebook's Five New Reaction Buttons: Data, Data, Data, Data, and Data," *Slate.com*, February 24, 2016, http://www.slate.com/blogs/future_tense/2016/02/24/facebook_s_5_new_reactions_buttons_are_all_about_data_data_data.html.

6. And I thank Kaling for the inspirational title. For a fuller dose of Mindy Kaling's humor, see her memoir in full: *Is Everybody Hanging Out Without Me? (And Other Concerns)* (New York: Three Rivers Press, 2012).

7. For more on FOMO, see Andrew Przybylski, Kou Murayama, Cody R. DeHaan, and Valerie Gladwell's "Motivational, Emotional, and Behavioral Correlates of Fear of Missing Out," *Computers in Human Behavior* 29, no. 4 (2013): 1841–1848. It turns out the media has gone pretty wild for articles about FOMO, too, and for a selection of them, see the following: Safronova's article "On Instagram, the Summer You Wish You Were Having," *New York Times*, August 20, 2015, D1–D7, in which she looks at FOMO as a side effect of social media sharing. Jenna Wortham's "Feel Like a Wallflower? Maybe It's Your Facebook Wall," *New York Times*, April 10, 2011, BU3, also presents FOMO as one of the negative consequences of the immediacy of information received through social media sites like Facebook, Twitter, Foursquare,

and Instagram. Wortham notes that the immediacy with which we can now receive updates about what others are doing serves to amplify the anxiety, inadequacy, and irritation one can feel when using social media. Hephzibah Anderson, in her article "Never Heard of Fomo? You're So Missing Out," *Guardian*, April 16, 2011, notes that the ability to instantaneously post about what one is doing causes others to feel like they are missing out and also prevents us from living in the moment and enjoying an experience that is ours alone. Finally, for a more positive take on the FOMO phenomenon, see Holly Williams, "Fear of Missing Out May Be a Latter-Day Anxiety, but We All Need to Embrace It," *Independent*, May 23, 2015, which—while acknowledging its potential negative impacts on our lives—argues that FOMO can also have positive effects, in that it can encourage us to go out and engage in activities we might otherwise not be willing to do.

8. In their article "Research Note—Why Following Friends Can Hurt You: An Exploratory Investigation of the Effects of Envy on Social Networking Sites among College-Age Users," *Information Systems Research* 26, no. 3 (2015): 585–605, Hanna Krasnova and colleagues look at survey responses from 1,193 college-age Facebook users to investigate the role of envy in the social networking site context as a potential contributor to undesirable outcomes such as depressive symptoms and anxiety. The authors' results showed that envy is associated with reduced cognitive and affective well-being as well as increased reactive self-enhancement. In "Friend Networking Sites and Their Relationship to Adolescents' Well-Being and Social Self-Esteem," *CyberPsychology & Behavior* 9, no. 5 (2006): 584–590, Patti M. Valkenburg, Jochen Peter, and Alexander P. Schouten describe a survey they administered to 881 adolescents, aged ten to nineteen, who had an online profile. The survey results showed that the frequency with which adolescents used the site had an indirect effect on their social self-esteem and well-being. The use of the friend networking site stimulated the number of relationships formed on the site, the frequency with which adolescents received feedback on their profiles, and the tone of this feedback. Positive feedback on their profiles enhanced adolescents' social self-esteem and well-being, whereas negative feedback decreased their self-esteem and well-being.

9. A total of 738 students responded to this optional survey question.

10. A total of 738 students responded to this optional survey question.

Chapter 2

1. In "First Thought, Worst Thought," *New Yorker*, January 13, 2014, Mark O'Connell muses on the act of writing something regrettable on social media, as well as on the swift consequences such a mistake can bring. One of the most publicized examples of this occurred in 2013, when Justine Sacco, a public relations executive, boarded a plane for South Africa and tweeted, "Going to Africa. Hope I don't get AIDS. Just kidding! I'm white." O'Connell notes that in the twelve hours she spent en route to Cape Town, Sacco became the unknowing subject of "a kind of ruinous flash-fame" as her tweet went viral, drawing anger and derision from thousands, which ultimately ended in her swift and public firing from her job. See also nonfiction author Jon Ronson's article about Justine Sacco, "How One Stupid Tweet Blew Up Justine Sacco's Life," *New York Times Sunday Magazine*, February 12, 2015. This article was excerpted from Ronson's book on the same subject: *So You've Been Publicly Shamed* (New York: Riverhead, 2015).

2. In her article "Beware: Potential Employers See the Dumb Thing You Do Online: The Spread of Social Media Has Given Hiring Companies a Whole New List of Gaffes to Look For," *Wall Street Journal*, Eastern Edition, October 29, 2012, B8, Leslie Kwoh details how employers are increasingly scouring social media to vet potential employees. Data show that in 2012, two in five companies used social networking sites like LinkedIn, Facebook, and Twitter to screen candidates. Social media "gaffes" that hiring managers reported as raising red flags included disparaging a former employer, using inappropriate language, offering excessive personal information, and exhibiting racist or sexist behavior, as well as any illicit activity such as drinking and driving or using illegal drugs. Kwoh notes that Millennials are especially vulnerable to these mistakes because they have a greater presence on social media and have grown up sharing their thoughts and feelings online. Then, in "10 Social Media Blunders That Cost a Millennial a Job—or Worse," *Time*, September 5, 2014, Susie Poppick reports that upwards of 93 percent of recruiters now check out social media profiles of prospective employees. Poppick provides a list of the ten most egregious mistakes job seekers can make on social media, including drinking in a photo, complaining about an old job,

making fun of clients or donors, and sexual oversharing. Also, in his article "The 7 Social Media Mistakes Most Likely to Cost You a Job," *Time*, October 16, 2014, Jacob Davidson looks at a 2014 survey by the recruiting platform Jobvite, which shows that 93 percent of hiring managers are now likely to review a candidate's social profile before making a hiring decision. The survey shows that 55 percent of hiring managers have taken a second look at a candidate based on what they find, with 61 percent of these second looks being negative. Davidson notes that job seekers should refrain from making references to illegal drugs, posting messages or photos of a sexual nature, using profanity, or posting about guns or alcohol. Posting one's political affiliation and having poor grammar were also turnoffs for many hiring managers. Aspects of social media profiles that hiring managers found favorable included information about volunteering or donating to charity. Finally, in a survey of 2,303 hiring managers and human resource professionals conducted in 2012, results showed that 37 percent of employers use social networks to screen potential job candidates. Of those employers that utilized social networks, 65 percent said they did so to see if potential employees presented themselves professionally, 51 percent wanted to know if the candidate was a good fit with the company culture, and 45 percent did so to learn more about the candidate's qualifications. Roughly a third (34 percent) of employers using social networks stated that information found on a candidate's social media profile caused them to not hire the candidate; specific red flags included inappropriate photos, evidence of drinking or drug use, poor communication skills, bad-mouthing former employers, and discriminatory comments related to gender, race, or religion. See Jacquelyn Smith, "How Social Media Can Help (or Hurt) You in Your Job Search," *Forbes*, April 16, 2013.

3. In his article "College Admissions Officials Turn to Facebook to Research Students," *U.S. News & World Report*, October 10, 2011, Ryan Lytle notes that college admissions officials are beginning to check social media profiles, such as applicants' Facebook pages, when making admissions decisions. In a Kaplan Test Prep survey of admissions officials at 359 colleges and universities, 12 percent of respondents noted that the use of vulgar language in a status update or depictions of alcohol consumption in photos negatively

impacted a prospective student's admissions chances. Lytle suggests that college-bound Facebook users not only be cognizant of the potentially negative aspects of their social media profiles but also take advantage of the fact that college admissions officers will be checking their profile by posting projects, research, and writing as a way of showcasing their strengths and achievements.

4. Michel Foucault, *Discipline and Punish*, trans. Alan Sheridan (New York: Vintage Books, 1975), 201. This connection between Foucault's "panopticism" and this project I owe to a serendipitously timed reading of *Discipline and Punish* and lecture about it by Ann Burlein, a philosophy professor colleague at Hofstra, for a course we were team-teaching in the Honors College.

5. See Daniel Trottier, *Social Media as Surveillance: Rethinking Visibility in a Converging World*, (London: Routledge, 2012). For more information on university officials scrutinizing their students' social media activity, see Daniel Trottier's article, "Mutual Transparency or Mundane Transgressions? Institutional Creeping on Facebook," *Surveillance and Society* Vol. 9, Issue ½ (2011): 17–30.

6. danah boyd, *It's Complicated: The Social Lives of Networked Teens*, (New Haven, CT: Yale Univerity Press), 54–59.

7. In "How to Clean Up Your Social Media for College Applications," *Huffington Post*, November 19, 2013, http://www.huffingtonpost.com/2013/11/19/social-media-college-applications_n_4303319.html, Megan Shuffleton discusses the need for Millennials who are preparing for college to clean up their social media profiles. Suggested steps to take include monitoring one's privacy settings, deleting any pictures that could be embarrassing or deemed inappropriate, looking over one's "liked" pages, and even changing one's username or Twitter handle to one that is more professional.

8. Howard Gardner and Katie Davis, *The App Generation: How Today's Youth Navigate Identity, Intimacy, and Imagination in a Digital World* (New Haven, CT: Yale University Press, 2013), 66. See pp. 61–76 for the entire discussion of the "polished self" and the "packaged self."

9. Ibid., 67.

10. This is the case, despite such famous political moments in history as the example of the Arab Spring. Social media played a key role in the Arab Spring uprisings that began in 2011. In his report "Social Media in the Arab World: Leading Up to the Uprisings of 2011" (Washington,

DC: Center for International Media Assistance, February 2, 2011), Jeffrey Ghannam details the awakening of free expression (largely via social media) that helped to break down the stranglehold of state-sponsored media and information monopolies in many Arab countries. While much of this free expression has been politically motivated, Ghannam notes that political and even more general forms of personal expression in many Arab countries have religious undertones, considering that hundreds of activists, writers, and journalists have been arrested and imprisoned for challenging government authority and insulting Islam. For more on the role of social media in the Arab Spring uprisings, see Philip N. Howard et al., "Opening Closed Regimes: What Was the Role of Social Media during the Arab Spring?," Social Science Research Network, 2011, and Habibul Haque Khondker, "Role of the New Media in the Arab Spring," *Globalizations* 8, no. 5 (2011): 675–679.

11. A total of 735 students responded to this optional survey question. Also, in a separate survey question about instruction and advice they've received from mentors, teachers, and parents about their behavior on social media, 123 of 568 respondents (23 percent) mentioned the importance of considering "future employers" every time they post as the main advice they received about using it. An additional 19 students said something to the effect of "Watch what you post if you ever want to get a job" as the *only* advice they received, and 10 other students said they were told (more or less) to always "keep things professional." Sixteen students mentioned that someone made sure to instill in them the knowledge that "I am a brand" (as one student put it), and 3 of these students went on to talk about "keeping things professional" and "future employers" in their same answers. This brings to 170 (30 percent) the total number of students who identified professionally related advice as the only social media advice they've been given.

Chapter 3

1. For more on the notion of social media as a space for "performance" and "exhibition" of self, see Xuan Zhao, Niloufar Salehi, Sasha Naranjit, Sara Alwaalan, Stephen Voida, and Dan Cosley, "The Many Faces of Facebook: Experiencing Social Media as Performance, Exhibition, and Personal archive," in *Proceedings of the SIGCHI*

Conference on Human Factors in Computing Systems, 1–10. ACM, 2013. See also José Van Dijck's fascinating exploration of "online self-presentation" through her comparative look at this subject on LinkedIn versus Facebook in her article, "'You Have One Identity': Performing the Self on Facebook and LinkedIn," *Media, Culture & Society* 35, no. 2 (2013): 199–215. And finally, see Zizi Papacharissi's chapter (12) on subject of "identity performance," "A Networked Self Identity Performance and Sociability on Social Network Sites," from Francis Lap Fung Lee, ed., *Frontiers in new media research*, Vol. 15, (London: Routledge, 2013), 207–221.

2. A total of 233 students chose to answer this essay question.

3. I've already mentioned Sherry Turkle's fascinating book on our experience of connectivity (or the lack of it) because of new technologies, but I find her latest book mentioned here in this chapter even more illuminating on this same subject and beyond. For anyone interested in social media and the way new technologies are changing our world, relationships, identities, and lives, both of Turkle's books in their entirety are essential reading, in my opinion. But for a look at Turkle's discussion of the ways we "edit" our speech because of technology in particular, see *Reclaiming Conversation: The Power of Talk in a Digital Age* (New York: Penguin, 2015), 22–23.

4. And it is, indeed, a business—a profitable one—for many young adults today. For an eye-opening look at this, and the notion of "microcelebrities" (the new "influencers" in our society who have a certain amount of social media fame," see Taffy Brodesser Akner's article "Turning Microcelebrity Into a Big Business," from *The New York Times Sunday Magazine* (September 19, 2014).

5. Take Essenia O'Neill, the Australian teen, social media star, and "name brand" who amassed nearly a million followers to her Instagram account, largely by posting attractive bikini photos and shots of herself in designer digs that highlighted her Barbie-like features and body. O'Neill became even more famous when the pressures to post and to look perfect and happy got to be too much and she retagged her photos with "the truth" about all the terrible things she was thinking and feeling while taking them, as well as posting teary videos explaining why she was doing away with all her accounts. For more on O'Neill, see Jonah Bromwich, "Essenia O'Neill, Instagram Star, Recaptures Her Life," *New York Times*, November 3, 2015.

Chapter 4

1. Many people agree with this student's assessment that lots of selfie-takers have gone too far. Making fun of selfies and those who take them has become a regular sport on television and in the media, especially those people whose selfies have become famous—or rather infamous—because of their extreme riskiness. We've all heard of those people who've done crazy (and crazy stupid) things like trying to take a selfie with a bear (and provoking the US Forest Service to release a statement warning tourists against doing just this) or taking a selfie during the Running of the Bulls in Pamplona (while actually running with the bulls), which is also prohibited by Spanish authorities for obvious reasons. For more on the bull-selfie, and the Spanish authorities' hunt for the culprit, see Jessica Durando, "Man Takes Selfie during Bull-Run Festival in Spain," *USA Today*, July 14, 2014, http://www.usatoday.com/story/news/nation-now/2014/07/13/selfie-runner-spain-festival-bulls/12594815/. Some people have died while taking selfies because they just can't resist taking that photo even in the most dangerous and precarious of situations—*a lot* of people apparently. See Jennifer Newton, "Selfies Kill More People Than Sharks as People Try to Impress Friends Online," *Daily Mail*, September 23, 2015, http://www.dailymail.co.uk/news/article-3244939/More-people-died-taking-selfies-killed-sharks-far-year-people-come-dangerous-way-impress-friends-online.html.

   For more on incidents in which taking selfies has resulted in injuries or even death, see Reuters, "Selfie Madness: Too Many Dying to Get the Picture," *New York Times*, September 3, 2015; Jessica Durando, "Police: Man Killed While Taking Instagram Selfie with Gun," *USA Today*, September 2, 2015, ; Jessica Mendoza, "Woman Hurt While Taking Photo with Bison: Why Can't People Resist Selfies?," *Christian Science Monitor*, July 26, 2015; and Kiran Moodley, "Couple Fall to Their Death Whilst Attempting Cliff Face Selfie," *Independent*, August 11, 2014.

2. In her book *Seeing Ourselves through Technology: How We Use Selfies, Blogs and Wearable Devices to See and Shape Ourselves* (New York: Palgrave Macmillan, 2014), Jill Walker Rettberg looks at how selfies, blogs, and other life-logging tools and applications have become important ways through which we understand ourselves.

Rettberg's analysis presents these tools and applications as three intertwined modes of self-representation: visual, written, and quantitative. For more on selfies as a form of identity performance, see Gabriel Fleur, "Sexting, Selfies, and Self-Harm: Young People, Social Media, and the Performance of Self-Development," *Media International Australia, Incorporating Culture & Policy* 151 (May 2014): 104–112. Here, Fleur argues that selfies are one of the many ways that young people consciously, visibly, and deliberately perform their identity online, and that social media and the structures of performative display are a way to reconceptualize youth and the relationship between social media and young people's self-development. Then, Haje Jan Kamps, in his book *Selfies: Self-Portrait Photography with Attitude* (Blue Ash, OH: How Books, 2014), celebrates the culture of social networking, in particular the art of taking selfies. In "Notable and Quotable: Selfie Hermeneutics," *Wall Street Journal*, August 17, 2015, A11, Tanya Abrams, Raul Alcantar, and Andrew Good discuss the University of Southern California's #SelfieClass, where freshmen students examine society's influence on self-identity and how selfies reflect and affect the global culture in which we live. The authors note that in class discussions and individual interviews, students admit that their selfies often reveal subconscious feelings about their own sexuality or ethnicity, and that they have used selfies to distance themselves from one group in the hopes of being accepted by another. Finally, in his article "Notes to Self: The Visual Culture of Selfies in the Age of Social Media," *Consumption Markets & Culture*, Taylor & Francis Online, July 3, 2015, 1–27, Derek Conrad Murray explores the cultural fascination with selfies with a specific interest in the self-imaging strategies of young women in their teens and early twenties. Murray explores the political urgency at the heart of the selfie phenomenon and contemplates whether the urge to compulsively produce and share self-images is mere narcissism or a politically oppositional and aesthetic form of resistance.

3. In "My Selfie, Myself," *New York Times*, October 20, 2013, 1–9, Jenna Wortham discusses the social and psychological aspects of self-portrait photographs, or selfies. Wortham points out that the popularity of selfies raises questions about vanity, narcissism, and our obsession with beauty and body image, but she also notes that

self-portrait photographs can often be more effective than text at conveying a feeling or reaction, and that receiving a photo of the face of someone we are talking to brings back the human element of the interaction—something critics often say is missing in electronic communications. See also John Suler's article "From Self-Portraits to Selfies," *International Journal of Applied Psychoanalytic Studies* 12 (June 2015): 175–180, where he looks at the evolution of the self-portrait and its democratization with digital technology.

4. In a study conducted by the Pew Research Center, survey research gathered in September 2009 showed that almost twice as many high school–aged girls use Twitter as their male counterparts (13 percent versus 7 percent, respectively). See Amanda Lenhart et al., "Social Media and Mobile Internet Use among Teens and Young Adults," *Pew Internet & American Life Project* (Washington, DC: Pew Research Center, 2010). In a more recent Pew Research Center study, conducted in 2012, survey research showed that of all Internet users, 71 percent of women use social networking sites, compared with only 62 percent of men. Overall, young adult women, aged eighteen to twenty-nine, were the most likely demographic group to frequent a social networking sites. Women were also more likely to use Facebook (72 percent versus 62 percent of men), Instagram (16 percent versus 10 percent of men), and Pinterest (25 percent versus only 5 percent of men), whereas men were slightly more likely to use Twitter (17 percent versus 15 percent of women). See Maeve Duggan and Joanna Brenner, "The Demographics of Social Media Users— 2012," *Pew Internet & American Life Project* (Washington, DC: Pew Research Center, 2013). Then, on a more popular level, in "The Social Media Gender Gap," *Bloomberg Business*, May 19, 2008, http://www. bloomberg.com/bw/stories/2008-05-19/the-social-media-gender-gapbusinessweek business-news-stock market-and-financial-advice, Auren Hoffman notes that while male and female young adults are just as likely to be members of social networking sites like Facebook, Myspace, and Flixster, young women are much more active on these sites than young men. For those over age thirty, the gender disparity widens, as men are found to barely be joining social networks, with the exception of LinkedIn. Adult women, however, are joining social networks in droves, with married women aged thirty-five to fifty being the fastest-growing segment. A suggested reason for this

gender disparity are that men view their Internet usage in a more transactional way than women, while women's Internet usage is more relationship-driven. In her article "Older Adolescents' Motivations for Social Network Site Use: The Influence of Gender, Group Identity, and Collective Self-Esteem," *CyberPsychology & Behavior* 12, no. 2 (2009): 209–213, Valerie Barker found that females were more likely than males to report high positive collective self-esteem, greater overall social network site usage, and higher use of social networking sites to communicate with peers. Conversely, males were more likely than their female counterparts to use social networking sites for social compensation and social identity gratifications.

5. For anyone interested in gender norms and stereotypes around girls during adolescence, *Meeting at the Crossroads: Women's Psychology and Girls' Development* by Lyn Mikel Brown and Carol Gilligan (Cambridge, MA: Harvard University Press, 1992) is a must—as is all the research and groundbreaking work on adolescent girls that follow it, especially Mikel Brown's more recent books on this subject, including *Packaging Girlhood: Rescuing Our Daughters from Marketers' Schemes*, co-authored with Sharon Lamb (New York: St. Martin's Press, 2006), as well as her companion analysis about adolescent boys, also co-authored with Sharon Lamb: *Packaging Boyhood: Saving Our Sons from Superheroes, Slackers, and Other Media Stereotypes* (New York: St. Martin's Press, 2009).

6. In their article "Face It: The Impact of Gender on Social Media Images," *Communication Quarterly* 60, no. 5 (2012): 588–607, Jessica Rose and her colleagues look at how gender is performed in the self-created digital images users upload to social networking sites like Facebook. Their results show that traditional gender stereotypes are upheld in these user-created images, with prominent male traits conforming to professional media depictions of men as active, dominant, and independent, while common female traits portrayed women as attractive and dependent. Also, in "Designing Gender in Social Media: Unpacking Interaction Design as a Carrier of Social Norms," *International Journal of Gender, Science and Technology* 6, no. 2 (2014): 223–241, Sofia Lundmark and Maria Normark focus on gender norms in relation to the design of interactive and digital products, services, and environments. In three separate case studies of different types of social media design and usage, the authors

find that gender norms influence the design of these interactive environments, and that the interface design in turn reinforces gender norms.

7. Please see, once again: Amanda Lenhart, et. al, "Teens, Social Media and Technology Overview" (Washington, DC: Pew Research Center, 2015).

8. There is a wide array of literature on how social media is affecting the body image of women and girls. For more on this subject, see Richard M. Perloff's article, "Social Media Effects on Young Women's Body Image Concerns: Theoretical Perspectives and an Agenda for Research," in *Sex Roles* 71, no. 11–12 (2014): 363–377. See also Renee Engeln-Maddox's article, "Cognitive Responses to Idealized Media Images of Women: The Relationship of Social Comparison and Critical Processing to Body Image Disturbance in College Women," from *Journal of Social and Clinical Psychology* 24, no. 8 (2005): 1114–1138. And also, Jasmine Fardouly, Phillippa C. Diedrichs, Lenny R. Vartanian, and Emma Halliwell's "Social Comparisons on Social Media: The Impact of Facebook on Young Women's Body Image Concerns and Mood," from *Body Image* 13 (2015): 38–45.

Chapter 5

1. This, despite the fact that so many studies about college students/ young adults and faith (including my own) find that incredibly high percentages of young adults—to the tune of around 80 percent— identify as spiritual and/or religious to at least some degree. For my work on this (that is unrelated to social media), see Donna Freitas, *Sex and the Soul: Juggling Sexuality, Spirituality, Romance and Religion on America's College Campuses* (New York: Oxford University Press, 2008). One of the best longitudinal studies around on young adults and spirituality/religion in general is Christian Smith's NSYR (National Study of Youth and Religion), with his publication *Soul Searching: The Religious and Spiritual Lives of American Teenagers* (New York: Oxford University Press, 2005) a must read for anyone interested in young adult religiosity. There are many additional significant publications by Christian Smith that have emerged from NSYR, *Souls in Transition: The Religious and Spiritual Lives of Emerging Adults* (New York: Oxford University

Press, 2009), *Lost in Transition: The Dark Side of Emerging Adulthood* (New York: Oxford University Press, 2011), and *Young Catholic America: Emerging Adults In, Out of, and Gone from the Church* (New York: Oxford University Press, 2014), among them. I also recommend Kenda Creasy Dean's important reading on this subject: *Almost Christian: What the Faith of Our Teenagers is Telling the American Church* (New York: Oxford University Press, 2010).

2. A total of 731 students chose to answer this question.

Also, with respect to young adult religious self-disclosure online specifically, Piotr S. Bobkowski and Lisa D. Pearce measured this in public Myspace profiles for a subsample of NSYR wave 3 respondents. The authors' findings show that 62 percent of profile owners identified their religious affiliations online, although only 30 percent said anything about religion outside of the religion-designated field (which is only slightly more than the 25 percent in my survey). Most reports of affiliation (the 80 percent mentioned above) were consistent with the profile owner's reported affiliation on the survey. See Piotr S. Bobkowski and Lisa D. Pearce, "Baring Their Souls in Online Profiles or Not? Religious Self-Disclosure in Social Media," *Journal for the Scientific Study of Religion* 50, no. 4 (2011): 744–762. Then, according to a report by the Pew Research Center, in an average week, 20 percent of Americans (overall) share their religious faith on social networking websites or apps, while 46 percent of US adults see someone else share their religious faith online. The Pew survey also showed that young adults aged eighteen to twenty-nine are about twice as likely as Americans aged fifty and older to see people sharing their faith online—though the students I interviewed certainly do not seem to support this finding. In the Pew study, evangelicals and black Protestants were far more likely than other major religious groups to say they shared their faith online. Finally, Americans who say they attend religious services frequently are more likely to say they engage in electronic forms of religious activity than those who attend services less often. See Pew Research Center, "Religion and Electronic Media" (Washington, DC: Pew Research Center, 2014). Also, the essays in *Social Media and Religious Change* (Berlin: De Gruyter, 2013), a collected volume edited by Marie Gillespie, David Herbert, and Anita Greenhill, address the interaction between social and mass media in the construction of contemporary religion and spirituality more generally. Topics discussed include the implications of social media for religious authority, the implications of mediatization for community relations, and the challenges of social media for traditionally bounded religious communities. And finally, see *Digital Religion, Social Media and Culture: Perspectives, Practices and Futures* (New York: Peter Lang, 2012) in its entirety, where Pauline Hope Cheong and colleagues collect and discuss current research on the complex interactions between religion and computer-mediated communication. Contributions center around the question of how core religious understandings of identity, community, and authority will be shaped and reshaped by the communicative possibilities of Web 2.0 (e.g., social networking sites, blogs, wikis, mobile apps).

3. In her article "Religion on the Go: Believers Embrace Social Media," *Sentinel*, July 2014, B1, Nicole Williams discusses how churches and churchgoers are using social media to connect with others and spread the good word. She describes the Bible app "YouVersion," which has more than 146 million downloads, provides users with 924 versions of the Bible in 628 different languages, and gives users the ability to share Bible verses with their social networks. Williams also gives examples of modern churches that have fully embraced social media, one of which has its own app and encourages worshipers to follow along with the service using their mobile app.

Chapter 6

1. Once again, see the following books in their entirety for an extensive treatment of these issues, and the ways we are learning to "edit" ourselves: Howard Gardner and Katie Davis, *The App Generation: How Today's Youth Navigate Identity, Intimacy, and Imagination in a Digital World*; Sherry Turkle, *Reclaiming Conversation: The Power of Talk in a Digital Age*.

2. In "This Text Will Self-Destruct; Snapchat Sweeps Campus," *New York Magazine*, October 22, 2012, http://nymag.com/news/intelligencer/snapchat-2012-10/, Robert Moor talks with students at a Connecticut high school about their use of Snapchat. While parents assume that the app's main use will be for sexting, students claim it is

actually used as a way to take and share funny, ugly, or weird selfies (what the students from my study absolutely confirm).

3. In "Social Networking App Yik Yak Is 'Outrageously Popular on College Campuses,'" *USA Today*, September 30, 2014, Dan Reimold looks at several college newspapers to examine how Yik Yak is being used and reacted to on college campuses, with concerns raised about how the ability to post anonymously often results in mean-spirited, derogatory comments toward others, but also an acknowledgment that this anonymity can empower those individuals who might otherwise remain silent to feel comfortable sharing their thoughts, feelings, and concerns. See also Jonathan Mahler, "Who Spewed That Abuse? Yik Yak Isn't Telling," *New York Times*, March 9, 2015, A1–B4, and Evelyn M. Rusli and Jeff Elder, "Behind App's Rise, Dark Side Looms," *Wall Street Journal*, November 26, 2014, B1. These articles look at how social media applications like Yik Yak are being used for cyberbullying by both teenagers and college-age students, including how the apps have been used to send threats of mass violence or sexual assault at several US.college campuses. Then, in "Campus Uproar over Yik Yak App after Sex Harassment, Murder." CNN, May 7, 2015, http://money.cnn.com/2015/05/07/technology/yik-yak-university-of-mary-washington/index.html, David Goldman examines the growing controversy over anonymous social networking apps in the wake of the April 2015 murder of a college student in Virginia who was harassed and threatened on Yik Yak. Goldman discusses the lack of accountability when it comes to responding to threats made on the app, as both the university and Yik Yak made little effort to address concerns voiced by a feminist group on campus about the sexist and threatening messages being posted about the group and its members prior to the murder. For more on the anonymity (or lack thereof) of such apps as Yik Yak, see also Geoffrey A. Fowler, "Past, Secrets You Share Online Aren't Always Safe," *Wall Street Journal*, Eastern Edition, February 26, 2014, D1–D3. Then, in "The Epidemic of Facelessness," *New York Times*, February 15, 2015, 1–7, Stephen Marche argues that anonymous and faceless online comments make the world more unethical. And on the subject of ethics, particularly (though not Yik Yak, since it did not yet exist), see Carrie James's *Disconnected: Youth, New Media, and the Ethics Gap* (Boston: MIT Press, 2014) for an extensive look

and analysis of how social media and new technologies are affecting moral and ethical decision-making in young adults.

4. To read about what happened at the University of Missouri and discussions of other Yik Yak–related campus threats, see Caitlin Dewey, "What Is Yik Yak, the App That Fielded Racist Threats at University of Missouri?" *Washington Post*, November 11, 2015.

5. For more on how and why college students are driving the trend of using anonymous social media apps, see Ellen Brait, "No Names Attached: College Students Drive Anonymous Apps Trend," *Guardian*, September 6, 2015. Brait notes that Yik Yak is currently used at more than two thousand colleges and universities worldwide. She argues that anonymous social media apps like Yik Yak and Whatsgoodly allow college students to talk about the drama in their lives and freely express their thoughts and opinions without the fear of having an ill-humored or insensitive comment traced back to them, thus allowing students to maintaining a positive digital footprint as they prepare to move into the professional world.

Chapter 7

1. Jenna Wortham looks at harassment on social media sites like Facebook and Twitter in "Trying to Swim in a Sea of Web Invective," *New York Times*, Sunday Business Section, December 14, 2014, B4. Wortham looks at recent court cases involving social media, harassment, and free speech and discusses the steps these social media sites are taking to curb cyberbullying. In his article "The Shamers and the Shamed," *New York Times*, April 30, 2015, D2, Nick Bilton discusses cyberbullying in relation to online shaming and the idea of the online mob mentality. The article also touches on notions of gender, pointing out that women are likely to be ridiculed on social media in ways that men do not often experience. Then, in Daniel B. Wood's article, "Cyberbullying: Should Schools Police Students' Social Media Accounts?," *Christian Science Monitor*, September 17, 2013, http://www.csmonitor.com/USA/Education/2013/0917/Cyberbullying-Should-schools-police-students-social-media-accounts-video, Wood discusses both support for and opposition to the Glendale (CA) Unified School District's hiring of a private firm to monitor the social media accounts of its fourteen thousand students

in the wake of two teen suicides in 2012. In Stephanie Rosenbloom's "Dealing with Digital Cruelty," *New York Times*, August 24, 2014, 1–7, she looks at how the anonymity of Internet comments encourages more uninhibited behavior by commenters and considers the psychological aspects of dwelling on negative feedback. For a comprehensive summary of the unique concerns and challenges that cyberbullying raises for children, parents, and educators, including the newest digital venues where cyberbullying is appearing and a look at occurrences of cyberbullying among adults and among children, see also Robin M. Kowalski, Susan P. Limber, and Patricia W. Agatston, *Cyberbullying: Bullying in the Digital Age*, 2nd ed. (Malden, MA: Wiley, 2012). Shaheen Shariff also provides an in-depth look at cyberbullying and suggests practical educational responses in *Cyberbullying: Issues and Solutions for the School, the Classroom, and the Home* (New York: Routledge, 2008). For more on the issue of freedom of speech on US college campuses amid the growing use of social media to bully students and professors, see Morton Schapiro, "The New Face of Campus Unrest," *Wall Street Journal*, Eastern Edition, March 19, 2015, A15.

2. For more on the It Gets Better project, check out the website: http://www.itgetsbetter.org. See also Lizette Alvarez, Lance Speere, and Alan Binder, "Girl's Suicide Points to Rise in Apps Used by Cyberbullies," *New York Times*, September 14, 2013, A1–A3, for their discussion of the link between cyberbullying and suicide. The authors write of the suicide of twelve-year-old middle school student Rebecca Ann Sedwick due to cyberbulling through the alleged use of online social media applications such as ask.fm, Kik Messenger, and Voxer. See also Sameer Hinduja and Justin W. Patchin, "Bullying, Cyberbullying, and Suicide," *Archives of Suicide Research* 14, no. 3 (2010): 201–221. Then, in their survey of 1,963 middle school children in one of the largest school districts in the United States, Hinduja and Patchin show that youths who experienced traditional bullying or cyberbullying, as either an offender or a victim, had more suicidal thoughts and were more likely to attempt suicide than those who had not experienced such forms of peer aggression. Victimization was also more strongly related to suicidal thoughts and behaviors than was offending.

3. In a survey of 276 adolescents aged fourteen to eighteen, Ozgur Erdur-Baker shows that 32 percent of students were victims of both

cyberbullying and traditional bullying, while 26 percent of students bullied others in both cyber and physical environments. Male students were more likely to be bullies and victims in both environments. Multivariate statistical analyses show that cyberbullying and traditional bullying are related for male students but not for female students. See Ozgur Erdur-Baker, "Cyberbullying and Its Correlation to Traditional Bullying, Gender and Frequent and Risky Usage of Internet-Mediated Communication Tools," *New Media & Society* 12, no. 1 (2010): 109–125. Also, in their survey of 2,186 middle school and high school students, Faye Mishna and colleagues show that more than 30 percent of respondents identified as being involved in cyberbullying as either victims or perpetrators, while one in four (25.7 percent) reported having been involved in cyberbullying as both bully and victim. Several risk factors were common among those involved in cyberbullying, including the number of hours per day students use the computer and giving passwords to friends. See Faye Mishna et al., "Risk Factors for Involvement in Cyber Bullying: Victims, Bullies and Bully-Victims," *Children and Youth Services Review* 34, no. 1 (2012): 63–70. Finally, in 2006, the National Crime Prevention Council commissioned a study, in conjunction with Harris Interactive, Inc., to explore the issue of cyberbullying among middle school and high school students in the United States. The survey had four objectives: (1) explore teens' experiences with cyberbullying; (2) understand teens' emotional and behavioral reactions to cyberbullying; (3) probe what teens think would be the most effective ways to prevent or put a stop to cyberbullying; and (4) determine how teens define cyberbullying and what other terms they use to describe it. Results of the study showed that of the 824 students surveyed, 46 percent had experienced some form of cyberbullying, females were more likely than males to have been victims of cyberbullying (57 percent versus 43 percent, respectively), and the likelihood of experiencing cyberbullying tended to increase slightly with age. See National Crime Prevention Council and Harris Interactive, "Teens and Cyberbullying: Executive Summary of a Report on Research," February 28, 2007.

4. For more on this study, see Sarah Konrath, Edward H. O'Brien, and Courtney Hsing, "Changes in Dispositional Empathy in American College Students over Time: A Meta-Analysis," *Personality and Social Psychology Review* 15, no. 2 (2011): 180–198.

5. For many teens, there is a fine line between bullying and plain old drama, with cyberbullying occupying another category altogether. To read more on these distinctions and the importance of this terminology (and the ways that adults today tend to lump everything under the heading of "bullying" and "cyberbullying," see Alice Marwick and danah boyd's article, "'It's just drama': teen perspectives on conflict and aggression in a networked era," in *Journal of Youth Studies* 17, no. 9 (2014): 1187–1204. In *It's Complicated*, boyd also spends a considerable amount of time discussing these categorical distinctions, as well as discussing the relationship between social media and adolescent cruelty/meanness. See boyd, *It's Complicated*, 128–52.

Chapter 8

1. See Ilana Gershon, *The Breakup 2.0: Disconnecting over New Media*. (Ithica: Cornell University, 2010) for an overview of how social media is affecting the way people experience breaking up.
2. For a look at how being public about one's relationship on Facebook can affect men's and women's levels of satisfaction (and dissatisfaction) in a relationship, see the article by Lauren M. Papp, Jennifer Danielewicz, and Crystal Cayemberg, "'Are We Facebook Official?' Implications of Dating Partners' Facebook Use and Profiles for Intimate Relationship Satisfaction," from *Cyberpsychology, Behavior, and Social Networking* 15, no. 2 (2012): 85–90.
3. In her article "The Limits of Friendship," *New Yorker*, October 7, 2014, Maria Konnikova discusses how social media is affecting the size of our social circles and the strength of the ties we keep with our "friends." Konnikova notes that while social networking sites like Facebook and Twitter allow us to connect with and be connected to more people than ever before, without investing in face-to-face time, we lack deeper connections to them, and the time we invest in superficial relationships comes at the expense of more profound ones.

Chapter 9

1. Many critics argue that mobile dating apps are changing (in a negative way) the very fabric of how we engage in courtship and develop

intimate relationships. In *Vanity Fair*, journalist Nancy Jo Sales discusses how hookup culture has collided with mobile dating apps like Tinder, Hinge, Clover, and How About We, which, she argues, are transforming how we (and especially young people) meet, develop relationships, and have sex. Tinder executives did not respond well to the article, unleashing a thirty-tweet rant in which they criticized Sales for not touching on the positive experiences the majority of the app's users have on a daily basis. Some critics view mobile dating apps in a more positive light, however, arguing that in addition to making hooking up easier, they have also leveled the playing field and that, for women, because of the ability to prescreen potential suitors, hooking up in the digital age is the safest and most liberating it's ever been. Regardless of one's take on these mobile dating apps, their popularity cannot be denied. Tinder, for instance, processes more than one billion swipes and matches approximately twelve million people each day. See Peggy Drexler, "Millennial Women Are Taking a Laissez-Faire Approach to Romance," *Huffington Post*, January 30, 2015, http://www.huffingtonpost.com/peggy-drexler/-millennial-women-are-tak_b_6578116.html; Molly Wood, "Led by Tinder, a Surge in Mobile Dating Apps," *New York Times*, February 4, 2015; Nancy Jo Sales, "Tinder and the Dawn of the 'Dating Apocalypse,'" *Vanity Fair*, September 2015; and Charles Riley and Hope King, "Tinder Says It 'Overreacted' to Vanity Fair Story with 30-Tweet Rant," CNN Money, August 12, 2015, http://money.cnn.com/2015/08/12/media/tinder-vanity-fair-twitter/index.html.

2. I write extensively about the nature of hookups, sex, and dating on college campuses in Donna Freitas, *Sex and the Soul: Juggling Sexuality, Spirituality, Romance and Religion on America's College Campuses* (New York: Oxford University Press, 2008) and also *The End of Sex: How Hookup Culture is Leaving a Generation Unhappy, Sexually Unfulfilled, and Confused about Intimacy* (New York: Basic Books, 2013).

3. In *Sexting: Gender and Teens* (Rotterdam: Sense Publishers, 2015), Judith Davidson provides a close-up look into the intimate and gendered world of teens and those who live with and work with them. Drawing on interviews with teens, parents, and caregivers, Davidson explores the new digital world that is still permeated by the beliefs and patterns of earlier patriarchal structures. Davidson's findings reveal that there are significant gendered differences among both teens and

adults in their perspectives on sexting. Also, as reported in their article "Prevalence and Characteristics of Youth Sexting: A National Study," *Pediatrics* 129, no. 1 (2012): 11–20, Kimberly J. Mitchell and colleagues conducted a cross-sectional national telephone survey of 1,560 youth Internet users aged ten to seventeen to determine their level of engagement with sexting. Their data revealed that 2.5 percent of youth had appeared in or created nude or nearly nude pictures or videos; however, when restricted to sexually explicit images, only 1.0 percent of respondents had engaged in this behavior. Of those surveyed, 7.1 percent stated they had received nude or nearly nude images of others, while 5.9 percent reported receiving sexually explicit images. In another survey, Donald S. Strassberg and colleagues interviewed 606 high school students and found that nearly 20 percent of respondents had sent a sexually explicit image of themselves via cell phone, and nearly twice as many had received such a photo from someone else. Of those who had received a sexually explicit photo, more than 25 percent had forwarded the photo to others. Nearly a third of students who had sent a sext did so despite believing there could be serious legal and other consequences attached to their actions. See Donald S. Strassberg et al., "Sexting by High School Students: An Exploratory and Descriptive Study," *Archives of Sexual Behavior* 42, no. 1 (2013): 15–21. Also, results from a Pew Research Center survey of youth aged twelve to seventeen conducted in 2009 showed that 4 percent of cell phone–owning teenagers have sent sexually suggestive nude or nearly nude images of themselves to someone else via text messaging, while 15 percent indicated that they have received such images from others. These percentages increase for older teens, with 8 percent of seventeen-year-olds sending such images and 30 percent having received sexually suggestive images. Teens who pay their own phone bills are nearly six times as likely to send sexually suggestive images as those teens who do not pay for or pay only a portion of their cell phone bill. Focus groups also revealed three main scenarios for sexting: (1) exchange of images solely between two romantic partners, (2) exchanges between partners that are shared with others outside the relationship, and (3) exchanges between people who are not yet in a relationship but where at least one person hopes to be. See Amanda Lenhart, "Teens and Sexting: How and Why Minor Teens Are Sending Sexually Suggestive Nude or Nearly Nude Images via Text Messaging," *Pew Internet & American*

Life Project (Washington, DC: Pew Research Center, 2009). And finally, Michelle Drouin and Carly Landgraff surveyed 744 college students regarding how texting and sexting impacted attachment in their romantic relationships. Results showed that texting and sexting are relatively common in young adult romantic relationships, and texting and sexting are both significantly related to attachment style. Texting was more common among those with secure attachments, while sexting was more common among those with insecure attachments. For those exhibiting attachment avoidance, men were more likely than women to send sexts to their relationship partners. See Michelle Drouin and Carly Landgraff, "Texting, Sexting, and Attachment in College Students' Romantic Relationships," *Computers in Human Behavior* 28, no. 2 (2012): 444–449.

Chapter 10

1. That smartphones and the ease with which they connect us to social media is affecting college students' study habits is not surprising. In "Distractions, Distractions: Does Instant Messaging Affect College Students' Performance on a Concurrent Reading Comprehension Task?," *CyberPsychology & Behavior* 12, no. 1 (2009): 51–53, Annie Beth Fox, Jonathan Rosen, and Mary Crawford show that students who instant-messaged while performing a reading comprehension task took significantly longer to complete the task than students who completed the reading comprehension assignment uninterrupted. Additional analyses also showed that the more time participants reported spending instant-messaging, the lower their reading comprehension scores and the lower their self-reported GPA. Also, in their article "Does Multitasking with Mobile Phones Affect Learning? A Review," *Computers in Human Behavior* 54 (2006): 34–42, Quan Chen and Zheng Yan show that mobile phone multitasking is prevalent among students. Moreover, this multitasking proves to be a distraction to learning in various ways. Chen and Zheng look at both how and why it impairs learning and offer suggestions for how to prevent mobile phone distraction.

2. And, as everyone knows by know, staring at one's smartphone in the middle of the night is about the worst thing we can do as far as getting our rest goes. In "How Smartphones Hurt Sleep," *Atlantic*,

February 24, 2015, http://www.theatlantic.com/health/archive/2015/ 02/how-smartphones-are-ruining-our-sleep/385792/, Olga Khazan discusses how, although smartphones may make our lives better in various ways, one way they may be hurting us is in the realm of sleep. Khazan points to a 2012 Time/Qualcomm poll of 4,700 individuals in seven countries, which showed that younger people were more likely to say that they don't sleep as well because they are connected to technology all the time. This could be due in part to the fact that smartphones and tablets emit blue light, which communicates to the brain that it's morning, whereas red light signals that it is time to sleep.

3. In "Millennials LOVE Their Smartphones: Deal with It," *USA Today*, September 27, 2014, http://www.usatoday.com/story/money/ personalfinance/2014/09/27/millennials-love-smartphones-mobile-study/16192777/, Lisa Kiplinger reports on a study by Zogby Analytics that found almost 90 percent of Millennials say their phones never leave their sides. For 80 percent of Millennials, the first thing they do when the wake up in the morning is reach for their smartphones, while 78 percent spend more than two hours per day texting, tweeting, talking, and searching online. Moreover, 47 percent of Millennials access businesses via their smartphones at least once a day, and 14 percent say they would not even do business with a company that doesn't have a mobile site or app.

4. That is 5% (31) out of a total of 608 students who chose to answer this optional question.

5. Sherry Turkle, *Reclaiming Conversation: The Power of Talk in a Digital Age* (New York: Penguin, 2015), 53.

Chapter 11

1. Students are not alone in quitting (often temporarily). It seems to be becoming a wider trend among all social media users. In her articles "Why I'm Quitting Social Media for 30 Days," *Huffington Post*, January 1, 2014, http://www.huffingtonpost.com/jordan-turgeon/ quit-social-media_b_4519567.html, and "I Quit Social Media (and I Don't Miss It Yet)," *Huffington Post*, January 23, 2014, http:// www.huffingtonpost.com/jordan-turgeon/quit-social-media_b_ 4655024.html, Jordan K. Turgeon discusses why she decided to quit

social media and what she has learned from the experience. After a few weeks away from social media, like many of the students who participated in this study, Turgeon notes that the experience has been an enjoyable one—her stress levels have dropped, she has more time for other things, and she is sleeping better. In his article "The Anti-social Network: Users Are 'Detoxing' from Facebook, Twitter," *MarketWatch*, August 1, 2015, http://www.marketwatch.com/story/the-anti-social-network-users-are-detoxing-from-facebook-twitter-2015-07-29?link=MW_home_latest_news, Charles Passy discusses the growing trend of people forgoing social media for the short or long term, noting that there is now a "National Day of Unplugging," a digital detox camp for adults, and in 2014 upwards of 16 percent of individuals who observed Lent planned to give up sites like Facebook and Twitter. Passy states that many users are beginning to see social media as a distraction that keeps them from going about the business of their day, while others see it as a toxic environment that creates feelings of envy and resentment. In "Swear Off Social Media, for Good or Just for Now," *New York Times*, July 2, 2014, Molly Wood discusses the different ways that one can "go dark" on social media, including deactivating or deleting accounts, as well as using apps like Xpire, DLTTR, and TweetDelete to limit one's digital presence. Wood notes that people who are choosing to abstain from social media often do so because they feel it is addictive, invasive, manipulative, and a distraction from more important tasks. Also, in "I'm Quitting Social Media to Learn What I Really Like," *Wired*, August 2, 2015, http://www.wired.com/2015/08/im-quitting-social-media-learn-actually-like/, Jessi Hempel discusses the pros and cons of how social media affects our day-to-day lives. Hempel notes that people are now engaging in digital detoxes, smartphone-free summer camps, and Facebook cleanses, citing as reasons the connection between social media use and anxiety, links between social media use and episodes of depression, and even legitimate questions about social media's impact on our brains.

2. A total of 295 students chose to answer this essay question.

3. In her article "Behind the 'Unlikes': Understanding Why People Quit Facebook," *Time*, September 19, 2013, http://healthland.time.com/2013/09/19/behind-the-unlikes-understanding-why-people-quit-facebook/, Maia Szalavitz discusses a study by Austrian researchers

who looked into why people choose to quit the social networking site Facebook. Their data showed that roughly half of those who left the site did so because of fears about privacy and ethical concerns about how their personal data would be used; others quit because they feared they were becoming addicted to the site. Szalavitz also mentions other studies that point to links between social media usage and low self-esteem, increases in depression, and FOMO, all of which can also cause people to shut down their social media accounts. Also, in "Why Facebook Makes You Feel Bad about Yourself," *Time*, January 24, 2013, http://healthland.time.com/2013/01/24/why-facebook-makes-you-feel-bad-about-yourself/, Alexandra Sifferlin discusses a study conducted by German researchers which showed that one in three people who visited Facebook felt worse after spending time on the site. This was especially true of visitors who viewed other people's vacation photos while on the site and of those who did not post any content of their own while on the site. The researchers also found that some of these dissatisfied people ended up reducing their use of the site, whereas others left it altogether. The most common causes of frustration and dissatisfaction for users came from comparing themselves socially to their peers and from having fewer comments and likes and less general feedback than their friends.

4. A total of 412 students chose to answer this optional essay question.

5. This reminds me of how college students are learning to prize ambivalence with regard to sex and their partners within hookup culture. I write about this extensively in *The End of Sex: How Hookup Culture is Leaving a Generation Unhappy, Sexually Unfulfilled, and Confused about Intimacy* (New York: Basic Books, 2013).

Conclusion

1. The idea that the Facebooks and Instagrams of today are like the grand new boulevards of still nascent modern cities came from a serendipitously timed reading of Marshall Berman's *All That Is Solid Melts Into Air: The Experience of Modernity* (New York: Penguin, 1982), and in particular, his chapter "Baudelaire: Modernism in the Streets," pp. 131–171.

Appendix

1. See Deborah L. Tolman and Mary Brydon-Miller, eds., *From Subjects to Subjectivities: A Handbook of Interpretive and Participatory Methods* (New York: NYU Press, 2001).
2. For more on this, see Christian Smith, *What Is a Person? Rethinking Humanity, Social Life, and the Moral Good from the Person Up* (Chicago: University of Chicago Press, 2011), 277–314.

Index

Campus Ministry programs, 278
casual sex. *See* hookups
ChatRoulette, 121–2
Clementi, Tyler, 146
Clover, 195
colleges and universities. *See also* sorority/
    fraternity life
  adjustment to, 304*n*10
  classroom bans on smartphones, 61, 271
  counseling centers and, 279–80
  hookup culture on, 197–9
  need for classroom dialogue on effects of
    social media, 60–2, 266–7, 276–7
  on professionalization of online brand,
    53–4, 57, 61
  recommendations for, 270–80
  review of social media of students/
    prospective students, 11–2, 47, 49, 274–5,
    309*n*3, 310*n*7
  teaching how to unplug, 277–9
  Wi-Fi-free spaces, 273–4
comparison trap, and likes/retweets, 16–42.
    See also student interviewees
  accepting fake friends, 32–3, 65–6
  alienation due to, 23
  authenticity/inauthenticity, 21–2,
    25–8, 38–9
  competitive nature of posting, 24, 25–8, 36
  conformity, 22
  dependence on being liked and sense of self,
    28–33, 305*nn*3–4
  gender differences and, 31–2, 37
  grading oneself, 19–20
  high traffic times, 35–9
  meaning attached to number of likes, 34–5
  negative affect on resilience, 25–6
  party photos, 25
  popularity principle (Van Dijck), 19,
    41, 305*n*3
  posting of jokes, 30–1
  postings as autobiography, 35–9
  reuploading of photos, 32
  self-esteem and, 17–20, 24, 38–9, 85, 151,
    158, 307*n*8
  summary conclusion, 39–42, 307*n*8
  wording of posts, 30
conceited, use of term, 82
craft/cultivate/curate (3Cs), 15, 75–80.
    See also branding of self; forced
    positivity
creeping on people, use of term, 193
*Culture of Connectivity: A Critical History of
    Social Media, The* (Van Dijck), 19

dating, rarity of, 197–8
dating apps. *See* Clover; Grindr; Hinge; Tinder
Davis, Katie, 49
define the relationship talk (DTR). *See* Facebook
    official status
digital detox, 328*n*1
Discipline and Punish (Foucault), 47
downvotes. *See* comparison trap, and likes/
    retweets; Reddit
duck face selfies, 99–100

emotional nuance, 171
empathy, 167
employers
  curation of postings and, 54, 55–7, 90
  religious expression in social media and,
    111–2, 123
  review of social media of job seekers, 46–51,
    57–8, 77–8, 182, 234, 308*n*2, 311*n*11
  sexual orientation and, 182
entitlement. *See* selfie generation
envy, survey on, 307*n*8
evangelization. *See* religious expression, in
    social media
exhaustion, 10, 15, 24

Facebook, 7–8, 251. *See also* bullying/cyberbullying;
    comparison trap, and likes/retweets
  cathartic forums on, 12
  as CNN of envy, 39
  comparison to Instagram, 109, 131, 178
  comparison to LiveJournal, 127
  comparison to Snapchat, 132–4
  deactivation vs. deletion of accounts, 179
  message update feature, 280–1
  100 likes benchmark, 34
  ranking of photos/posts, 36
  reaction buttons, 306*n*5
  selfies on, 86
  social media résumés on, 49–51, 131
  30 likes benchmark, 34–5
  Timeline feature, 35–6
Facebook, professionalization of, 43–62
  Cleanups, 46, 48–9, 51–3, 61, 180
  economic background and, 53–8
  forced positivity, 44–6
  knowledge as power, 58–62
  lack of privacy, 47–8
  political/social posts, 44, 46, 49–51, 58–62
  potential employers and, 46–51, 54, 55–7,
    308*n*2, 311*n*11
  as reformed behavior, 53–8
  as virtual panopticism, 47

sorority/fraternity life
    applicant criteria and postings, 5
    guidelines/monitoring of postings, 5, 8
    peer enforcers in, 45
    politics, as off-limits for posting, 4
    pressure to conform, 1–9
    social probation, 8
spiritual practices, and unplugging, 277–8
sports, and device-free expectations, 272
streaking, 254
student interviewees
    Aamir, 44–6, 51, 53
    Adam, 87–8, 172–4
    Ainsley, 176–9
    Alex, 140
    Alima, 112–7
    Amy, 89–90, 149, 214, 240, 242
    Angela, 129–31, 139–40
    Avery, 38–9, 52, 214–5
    Blair, 210–1
    Bo, 59
    Brandy, 73–4, 75, 147–8, 194–5
    Brenda, 205
    Cherese, 63–8, 78, 92–3, 106, 219–21
    Corban, 143–5
    Daphne, 224
    David, 91–2
    Dinah, 118–9, 122, 184–7
    Eddison, 180–3, 184
    Elise, 83–4, 86, 87, 88, 94, 147, 235–7, 238
    Emily, 216–7
    Emma, 1–9
    Ephraim, 119–22
    Fara, 78–9
    George, 50–1
    Gina, 68–9, 212–3
    Grace, 124–9
    Hae, 241–3
    Hailey, 155–9
    Hannah, 26–8, 74–5, 188–9
    Ian, 97–8, 163–7, 171
    Jack, 159–63, 171
    Jackson, 86, 223–4
    Jae, 108–9
    Jake, 174–6
    Javier, 239–40
    Jennifer, 105–8, 110, 112
    Jeremy, 205–7
    Joe, 99–101
    John, 77–8
    Jose, 110–2
    Joy, 196, 204–5
    Justin, 136–7

    Katie, 243–5
    Kristin, 213
    Laura, 23–4, 96–7
    Lauren, 193–4, 232–5
    Lin, 59–60
    Lucy, 69–71, 72
    Mack, 54–5
    Mae, 149–55, 238
    Marcus, 226–7
    Margaret, 17–20, 305n3
    Maria, 36–8, 169–70
    Mark, 138–9, 202
    Matthew, 24–6, 35–6, 94, 132–4,
        202–3, 221–2
    Max, 90–1, 147
    May, 59
    Mercedes, 34–5, 216
    Michael, 16, 20–3
    Ming, 77
    Nikki, 55–6
    Nora, 168, 171
    Peter, 176–9
    Rob, 28–33, 305n3
    Sage, 196–7
    Sarah, 148
    Sheena, 56–7
    Stacie, 225
    Stephanie, 227–8
    Susan, 129
    Tanuja, 82, 86
    Tara, 98–9, 183–4
    Vidya, 203–4
    Zachary, 117–8, 119
suicide, 146, 168, 320n2
survey results, 11. *See also* methodology
    on anonymity, 137–8
    on being always "on call," 218–9
    on branding, 80
    on comparison trap, 40–1
    on concerns about potential employer
        reviews, 51, 311n11
    on curation of photos, 71–2
    essay questions, 148–9, 246, 247
    on expression of emotions, 126
    on forced positivity, 13
    gender of respondents, 95–6, 95f
    on limiting social media usage, 238
    on political/religious opinions, 110, 316n2
    on selfies, 84–5, 88
    on sexting, 207
    on taking breaks from phones, 215, 217
    on temporarily quitting social media, 238
    on use of Tinder, 197, 199

TED Talks, xvi
thin vs. thick skin, 159, 168–71, 257
Tinder, 194
    embarrassment about, 196
    flirting on, 197–8
    for hookups, 195–202, 324n1
    lesbian use of, 200
    negative views on, 200–1
    pros/cons of, 195–201
    sexting on, 206
    use of GPS on phone, 135, 195, 197
trolling, 159–63, 167. *See also* bullying/
    cyberbullying
Trottier, Daniel, 47
Tumblr, 129
Turkle, Sherry, 76, 229, 305n2
Twenge, Jean, 82
Twitter. *See also* comparison trap, and likes/
    retweets
    anonymity and, 129–31
    as autobiography, 35
    gender stereotypes and, 94
    political/social posts, 44, 46

unplugging, 210–8, 226–8, 263–4, 270–3,
    277–9. *See also* living in the moment;
    social media, avoidance/abandonment of
upvotes. *See* comparison trap, and likes/
    retweets; Reddit; Yik Yak

Van Dijck, José, 19, 41, 305n3
video games, 95
virtues for social media age, 256–64. *See also*
    specific virtues

abandonment of social media, 264
authenticity, 258–9
forgetting and letting memories
    fade, 260–1
living in the moment, 261–2
outlets for play, 262–3
unplugging, 263–4
vulnerability, 257
vulnerability
    cyberbullying as preying on, 15, 19–20,
        152–4, 157, 168
    as virtue for social media age, 257

Whatsgoodly , 321n5
Wi-Fi-free spaces, 273–4
willpower, 270–1, 273
wording of posts, 30, 67, 76, 133–4
work hard/play hard mentality, 12

Yik Yak, 7–8, 58, 251
    anonymity and, 12
    authenticity/inauthenticity, 258
    being liked on, 29
    as conducive to bullying, 146, 148–9,
        166, 169
    dark side of anonymity on, 135–42, 258,
        259, 275
    media coverage of, 135, 320n3
    as outlet for play, 263
    slut-shaming on, 139–40
    use of GPS on phone, 135
YouTube
    individual channels as branding, 79–80
    "It Gets Better" video, 146